Conservation and Sustainable Use

Techniques in Ecology and Conservation Series

Series Editor: William J. Sutherland

Bird Ecology and Conservation: A Handbook of Techniques
William J. Sutherland, Ian Newton, and Rhys E. Green

Conservation Education and Outreach Techniques
Susan K. Jacobson, Mallory D. McDuff, and Martha C. Monroe

Forest Ecology and Conservation
Adrian C. Newton

Habitat Management for Conservation: A Handbook of Techniques
Malcolm Ausden

Conservation and Sustainable Use: A Handbook of Techniques
E.J. Milner-Gulland and Marcus Rowcliffe

Conservation and Sustainable Use

A Handbook of Techniques

E.J. Milner-Gulland

and

Marcus Rowcliffe

OXFORD

UNIVERSITY PRESS

Great Clarendon Street, Oxford OX2 6DP

Oxford University Press is a department of the University of Oxford.
It furthers the University's objective of excellence in research, scholarship,
and education by publishing worldwide in

Oxford New York

Auckland Cape Town Dar es Salaam Hong Kong Karachi
Kuala Lumpur Madrid Melbourne Mexico City Nairobi
New Delhi Shanghai Taipei Toronto

With offices in

Argentina Austria Brazil Chile Czech Republic France Greece
Guatemala Hungary Italy Japan Poland Portugal Singapore
South Korea Switzerland Thailand Turkey Ukraine Vietnam

Oxford is a registered trade mark of Oxford University Press
in the UK and in certain other countries

Published in the United States
by Oxford University Press Inc., New York

© Oxford University Press 2007

British Library Cataloguing in Publication Data

Data available

Library of Congress Cataloging in Publication Data

Data available

Typeset by Newgen Imaging Systems (P) Ltd., Chennai, India
Printed in Great Britain
on acid-free paper by
Biddles Ltd., King's Lynn

ISBN 978–0–19–853036–7 (Hbk.)   978–0–19–853035–0 (Pbk.)

10 9 8 7 6 5 4 3 2 1

Acknowledgements

This book would not have been possible without the help and support of many people. We thank Bill Sutherland for convincing us to write it and supporting us throughout, and Ian Sherman and the team at OUP, particularly Stefanie Gehrig and Helen Eaton, for their encouragement and hard work.

EJMG thanks Hugh Possingham and the University of Queensland's Ecology Centre for hosting her while she was writing part of the book. Our employers Imperial College London and the Institute of Zoology supported us throughout. We thank all our colleagues and friends who have generously given us advice, pictures and information about their conservation work for our case studies. Six readers ploughed through the manuscript and made incredibly useful comments—thank you Matt, Kerry, Tom, Ambika, Pat and Nick. Our families have stuck with us and kept us going throughout the long difficult process—deepest gratitude to Martin and Rowan, and to Jo.

The main sources of inspiration for this book are the members of our research groups; PhD and MSc students and post-docs, past and present. Their commitment and determination to work for what they believe in, in very challenging circumstances, fills us with pride and admiration. These are people who will make a difference in the world. This book is dedicated to our students, with heartfelt thanks for all they have taught us, and in the hope that it will be useful to those who follow in their footsteps.

Contents

1

Introduction

1.1 Who is this book for?

This book is a practical handbook setting out the methods needed to conserve wildlife that is hunted or harvested. We are aiming at anyone who is interested in carrying out scientifically based conservation of exploited species. This includes managers who need to interpret the information available to them to decide on conservation actions, and those who carry out research to gain understanding of a problem and then act on that knowledge. It also includes people who are primarily carrying out scientific studies, but with the hope of making usable recommendations, for example, students who are working on the conservation of an exploited species for an MSc or PhD thesis.

We hope to provide a toolbox that covers the many aspects of conserving exploited species. The bedrock of effective conservation is to understand the biology of the exploited species, its habitat requirements and the effects of threatening factors on its population dynamics. If we are not correct about what is it that is causing a population to decline, then all other actions will be in vain. On the other hand, biological understanding is just the first step on the long road towards actually doing something to reverse declines. This requires an understanding of the **human setting** within which the threatening process is happening; who is involved, what their motives are and how the culture and institutions within which their activities occur drive their exploitation patterns. Action is even more difficult, as it requires engagement with people and institutions, so that **patterns of behaviour** can be changed; action needs to be at multiple scales, from passing laws to changing the perceptions of people in daily contact with wildlife.

We will look at techniques for getting information about the exploitation process, covering biological, social, economic and institutional angles. We then discuss how this information can be analysed to produce a scientifically based understanding of the dynamics of the system. Finally we discuss some of the considerations involved in translating this understanding into action. Overarching all of these is the issue of **uncertainty**—how do we make sound decisions when we don't have perfect knowledge, and how do we monitor the situation so that we can improve our understanding and pick up the warning signs of things going wrong?

We have to limit the range of issues we can consider. We focus here on species that are killed by humans for gain, for example, for subsistence food use or the international trade in traditional medicines. When we use the words use, exploitation, hunting, harvesting, then unless otherwise stated we are treating them as synonyms meaning killing an individual for gain. But we do not consider large-scale commercial harvests, such as are common in the forestry and fisheries sector, nor human-manipulated ecosystems such as plantations or fields. Instead we focus on the conservation of species that are hunted in their natural habitat, and at a relatively **small scale**. Our personal taxonomic bias is towards terrestrial vertebrates, but most of what we say is equally applicable to other living organisms that are similarly harvested; plants, corals, fish, fungi . . . It is also to some extent applicable to commercial operations and to use that does not kill (e.g. harvesting plant parts). Where methods are not broadly applicable, we point this out.

We necessarily take a **population-level approach**, because hunting is targeted at populations. Even when it is relatively non-selective, there is a limited taxonomic range taken in any one exploitation operation. This is another reason why we exclude commercial forestry and fisheries, some components of which, such as clear-cutting and bottom-trawling, destroy entire ecosystems. Of course even targeted hunting can have profound ecosystem-level effects, but we focus predominantly on studies that are concerned with the target organism itself. On the social side we focus our attention at the level of the individual harvester and their community, rather than at the national and international levels. Inevitably this means that some issues are treated in a cursory manner, but we do point out where processes at different scales have an important role to play in local sustainability.

Ignoring large-scale commercial operations does not imply that they are unimportant ecologically or socially. Commercial logging was a major factor in the 14% reduction in tropical forest area between 1990 and 2000 (FAO 2001), the collapse of the Grand Banks fishery had profound impacts on the local economy (Ruitenbeek 2001), and the over-exploitation of marine megafauna in the last few centuries has altered the ecology of the Caribbean beyond recognition (Pandolfi *et al.* 2001). But by focusing on the smaller scale, we aim to fill a gap. By and large, commercial forestry and fisheries operations are overseen by professional managers and scientists. In these systems, management may fail and science may confront profound uncertainties, but the focus is strongly on the species as a resource to be managed. Conversely, in situations where there is a conservation problem, there are generally no resource managers overseeing operations. Instead **conservationists** must diagnose the problem and devise methods for improving the situation. These are the people to whom this book is addressed.

1.2 Assessing threats and sustainability

1.2.1 Is exploitation the threat?

Because you are reading this book the assumption is that the species you are interested in is exploited. However, it is a further step from observing that exploitation is taking place to diagnosing it as the key threat to be tackled. Caughley and Gunn

(1995) give many examples of cases in which conservationists are quick to diagnose a threat, only to find that their interventions are unsuccessful because the obvious threat was not the true cause of a population's decline. For example, the decline of the large blue butterfly in the UK was at first attributed to collecting, then to habitat changes. It was only when its parasitism of a particular species of ant was discovered that the true cause of the decline was known and appropriate action could be taken (Elmes and Thomas 1992; Caughley and Gunn 1995).

The 'Bushmeat Crisis' is a major conservation issue (BCTF n.d.). There is much concern that hunting wildlife for food, particularly in the forests of West/Central Africa, is causing whole faunas to be wiped out, leaving empty forests behind them. On the broad scale, it is undoubtedly true that hunting is unsustainable over much of West and Central Africa (Fa *et al.* 2003). But when deciding on conservation priorities at the local scale this may not be true. When people have been eating their local wildlife for centuries, it may be that their use has reached sustainable levels. This is likely to be the case for Takoradi market, Ghana (Cowlishaw *et al.* 2005a). Bushmeat also illustrates the need to disentangle animal welfare concerns and cultural differences in attitudes to wildlife from the issue of whether hunting is actually a threat to population survival in a particular location.

The first two steps in conserving an exploited species are to decide first whether and why the species is of conservation concern; and second whether an intervention that aims to reduce hunting of that species actually is the best approach to addressing this concern. The most widely accepted method to judge whether a species is of conservation concern is its IUCN—World Conservation Union red list status (IUCN-SSC 2006). The IUCN red lists categorise species according to their risk of extinction, based on criteria such as the rate of population decline, small population size, limited range area and fragmentation. The most usual (but not the only) reason why exploited species are placed on the IUCN red list is rapid population decline. However, there are other valid reasons why a conservationist may be concerned about the impact of exploitation on a species. These include its cultural or economic significance, its importance as a component of the ecosystem, or its vulnerability to future over-exploitation. In these cases, the population may not currently be declining but an intervention is still appropriate.

Once we know why we are concerned about the species, we then need to assess the relative importance of exploitation compared to the plethora of other potential threats (such as habitat loss, hybridisation, alien invasives, disease). Rarely does a single factor act alone to cause extinction. Even if hunting is the main cause of decline, other factors are likely to come into play as the population becomes more threatened. Which factor should be tackled first is a product of the urgency of the problem, its seriousness and the cost-effectiveness of measures that could be taken. For example, Damania *et al.* (2003) used a model to suggest that the main threat to tiger populations in Indian Protected Areas is not direct killing of the tigers for sale, but depletion of their prey base for crop protection. Tigers have a comparatively high population growth rate and can withstand a fairly high level of hunting, so long as they have adequate prey to sustain the population.

Damania *et al.* argue that the focus of much conservation effort is on reducing tiger poaching rather than the potentially more effective strategy of reducing poaching of the tiger's prey. This contrasts sharply with the view of the Wildlife Protection Society of India that tiger poaching is driving the species rapidly to extinction (EIA-WPSI 2006). Despite the need to assess rigorously all the factors threatening a population, in many cases the issue that needs tackling is clear. For example, O'Brien *et al.* (2003) showed that despite other factors, particularly habitat loss, the Madagascan radiated tortoise is declining in range and population size principally because of high levels of exploitation to supply urban demand for tortoise meat.

Having decided why we are interested in a particular population, and what threats are acting on it, we next have to decide what outcomes we would like to see. The concrete objectives of a conservation intervention may include reversing a population decline, safeguarding a particular area for conservation or changing local attitudes towards a species. But the most usual objective cited for interventions to conserve exploited species is to ensure 'sustainable use'.

1.2.2 Defining sustainable use

The meaning of sustainable use is often not clearly defined by conservationists, because it is a difficult concept (Hutton and Leader-Williams 2003). It is useful to think about it as having three main components—biological, social and financial sustainability (Sample and Sedjo 1996). **Biological sustainability** implies that the activity does not compromise the integrity of biological systems—in the case of hunting a single species, this might translate into the population staying at a density high enough to ensure that it and the components of the ecosystem that it influences can persist into the long term. **Social sustainability** requires cultural appropriateness, social support and institutions that can function into the long term, and **financial sustainability** that the activity outcompetes unsustainable alternative activities in profit-generation. A simple and widely accepted definition, which is broad enough to encompass these aspects is the one used in Article 2 of the Convention on Biological Diversity:

Sustainable use means the use of the components of biological diversity in a way and at a rate that does not lead to the long-term decline of biological diversity, thereby maintaining its potential to meet the needs and aspirations of present and future generations. (CBD 1993)

The first part of this definition refers to biological sustainability, while social and financial sustainability are required in order to fulfil the second part. Hence biological sustainability is necessary but not sufficient for long-term sustainability of a conservation activity. An alternative broad-brush definition is that of the Brundtland Report (WCED 1987), which defined sustainable development as development that 'meets the needs of the present without compromising the ability of future generations to meet their own needs'. Sustainable use of natural resources is one aspect of sustainable development, and the definition works as well for use as for development more broadly.

Sustainability is partly a difficult concept because hunting systems are **dynamic**—they change over time, rather than being at a single equilibrium position. This is because there is variability in the system—the weather, chance events like accidents, political changes and so on. This variability knocks the system out of equilibrium, and acts at different scales in time and space. Table 1.1 gives some examples of the kinds of processes acting on sustainability at different scales. Some of these involve predictable and non-trending variation (e.g. seasonal food availability), some are not so predictable (short-term weather variation). Some involve sudden shifts in the system (e.g. wars), others relatively deterministic trends (e.g. human population growth). Sometimes, events do not have the expected effects on sustainability. For example, the Critically Endangered Virunga mountain gorilla population actually increased during the 1990s, a period of unparalleled civil unrest. One of the reasons for this is that the conservation programme put in place in peacetime was robust enough, and had built enough local support, to continue even when the rest of civil society was in disarray (Kalpers *et al.* 2003).

Given the range of scales at which these effects act, and their ubiquity, it is clear that assuming that human-environmental systems are at equilibrium is often

Table 1.1 Examples of social, financial and biological events that can affect sustainability of exploitation at different time-scales and spatial scales.

	Few days	Few months	Few years	Many years
Few km²	Cold/wet weather. Hunter health. Village festivals.	Food availability for prey. Alternative activities for hunters.	Village head changes. New job opportunities. Human and prey population size changes.	Change of culture and social institutions. Local prey adaptations.
Few hundred km²	Major regime change.	Fires, Droughts, Harsh winters, Floods,	Food preferences in city change. Changes in national government.	Evolutionary change.
Continental/ Global	Commodity price changes. Internet information transfer.	El Nino effects, e.g. coral bleaching. Reactions to commodity prices.	Wars. Environmental treaties.	Climate change.

going to be very unrealistic. It can also be extremely difficult to disentangle the effects of different processes acting at different scales, and hence to assess how best to intervene to move the system to a sustainable state. Sustainability, then, is more of a process than an equilibrium—it means that the system is able to maintain itself and to adjust to shocks (in biological language, how resistant and resilient it is).

This complexity means that when we say that exploitation is 'sustainable', we need to surround this statement with caveats. We need to define the temporal and spatial scale over which we are talking, and we need to acknowledge that the diagnosis of sustainability holds into the future only inasmuch as we are able to predict future changes. We can also assess how 'future-proof' the system is—how able it is to adapt to any changes that might occur; this is also an important component of sustainability (see Chapter 6 on the development of robust adaptive management strategies). This means that our sustainability assessments need a number of components. For example, if we were assessing the sustainability of fishing on a reef near a small village, we might ask:

Is the fished population roughly stable, and for how long has this been the case? [biological]
Is the ecosystem as a whole in good health? [biological]
Is the management of the fishery accepted by the community? [social]
Are fishermen making a good living from the fishery? [financial]
Is the system able to withstand any threats that are on the horizon? [future-proofing]

First, however, we need a framework for understanding the dynamics of the bio-economic system whose sustainability we are trying to assess.

1.3 The dynamics of bio-economic systems

In this section we give a brief overview of the underlying theory of harvesting, from the biological and economic perspectives. This is needed in order to understand some of the material in Chapters 2–4, particularly the section on biological reference points in Chapter 4. We come back to the theory and practice of modelling as a tool in the conservation of exploited species in Chapter 5.

1.3.1 The essentials of population dynamics

Some species can remain abundant under heavy exploitation, while others disappear under even the lightest harvesting. The explanation for this variation lies in the feedback between population size and population growth. Termed **density dependence**, this feedback is a central principle in population dynamics in general, and in the biology of sustainable use in particular. In a population regulated by density dependence, per capita growth rate (the net number of new individuals entering the population per existing individual per unit time) declines with increasing density. For example, under crowded conditions, food resources might be depleted, predation might intensify, or disease might take a bigger toll, any of which could cause increased mortality rates or reduced reproduction. Regardless of the mechanism of density dependence, the net result is a declining per capita population growth rate with increasing density.

The implications of density dependence for exploitation can be understood by imagining what happens when a population first becomes subject to a regular harvest. At its un-harvested equilibrium, births balance deaths and the net growth of the population is zero. The first harvest reduces the population below its natural equilibrium, internal competition eases and net growth becomes positive, resulting in a partial recovery by the time the next round of harvest starts. As the population falls, successive harvests become smaller while net growth increases. Finally, a new equilibrium is reached at which the harvest is exactly balanced by growth, and we have, in theory, a biologically sustainable harvest. Of course, a sustainable equilibrium is only achievable if the offtake does not exceed a certain limit, which we can define as a reference point.

1.3.1.1 The logistic model

The logistic is the simplest model of density dependent population growth. Being easy to analyse and understand, it has been used as the basis of much theory of sustainable use. This theory forms important background for understanding how models can be used to define sustainability benchmarks, and we therefore provide a brief introduction here. For more complete coverage of harvesting theory, see Clark (1990), Getz and Haight (1989) and Milner-Gulland and Mace (1998). Despite its simplicity the logistic model can also be used to model specific systems, and the theory can therefore be translated directly into practical applications in some cases.

The logistic is characterised by a linear decline in per capita growth rate with increasing population density (Figure 1.1(a)). At a very low population size this rate is maximal, while at the other end of the scale equilibrium population size (or carrying capacity) is defined by the point at which the growth rate falls to zero. Carrying capacity is commonly denoted K, while the maximum per capita growth rate is commonly denoted r_{max}, and these two parameters alone define logistic growth. Net growth in the population is the product of population size and per capita growth, and has a domed relationship with population size (Figure 1.1(b)). At small population size, although per capita growth is high, the small size of the population leads to little overall growth. Conversely, at large population size, low per capita growth leads to low net growth, despite the large population. Maximum net growth occurs at intermediate population size. The maximum sustainable yield (MSY) is equal to the net growth at this maximum. Assuming that harvest is directly proportional to harvester effort, we can also express equilibrium growth, and hence yield, as a function of a consistently applied effort. In this case equilibrium population size is linearly and inversely related to effort, and the yield–effort curve is therefore also domed (Figure 1.1(c)).

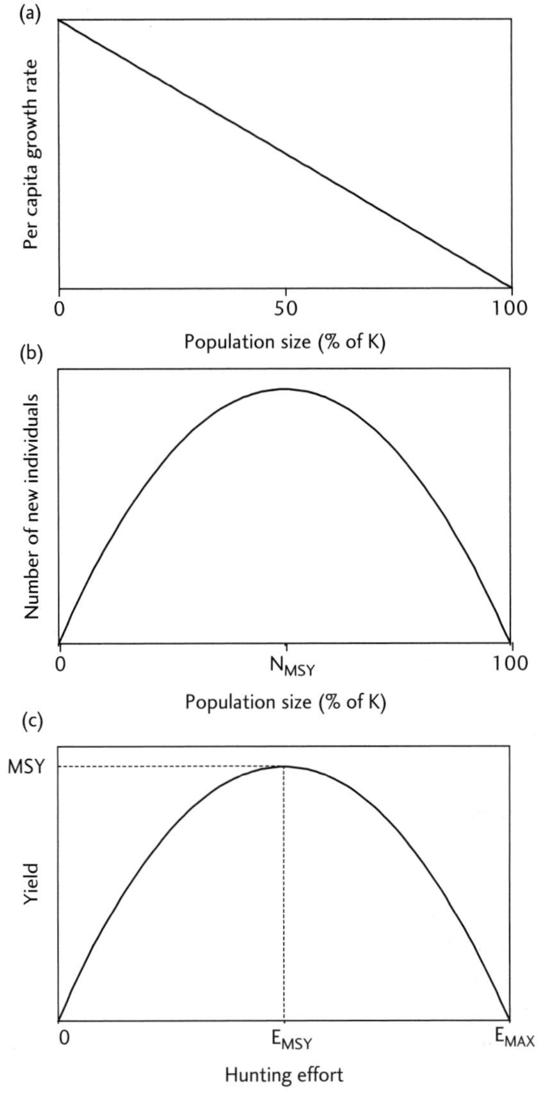

Fig. 1.1 The logistic growth model. a) The per capita growth rate (number of new individuals added to the population for each existing individual) as a function of population size; b) The total number of new individuals added to the population as a function of population size; c) Yield as a function of hunting effort, with the maximum sustainable yield shown.

1.3.2 Economics of supply and demand

In economic systems, the amount of a good in the market and its price are determined by the interplay between supply (the amount of a good a producer is willing to sell at a given price) and demand (the amount of a good a consumer is willing to buy

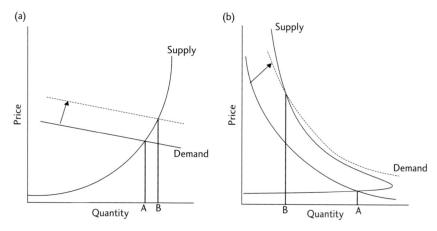

Fig. 1.2 Supply and demand for two contrasting natural resource systems. a) A sole owner supplies only up to the maximum sustainable yield level, and demand is elastic. This is a robust system in which an increase in demand, due to increased consumer income for example, leads to a new slightly higher, but still sustainable, equilibrium quantity supplied, at a slightly higher price. b) An open access system in which demand is highly inelastic. The supply curve is backwards-bending due to population depletion, and an increase in consumer income from the same equilibrium point A leads to a much lower quantity supplied from a depleted population, at a much higher price.

at a given price; Figure 1.2). Demand is determined by consumer tastes, their income and the presence of alternative goods. Supply is determined by producer costs, which in harvesting systems are mostly to do with the cost of finding and killing prey, and so are related to the prey's biology, for example, to population size and distribution patterns.

The slope of the supply and demand curves is called the **elasticity**, and this value is fundamental to determining how changes in external factors affect change in the equilibrium price and quantity (Begg *et al.* 2005). For example, if a good is a necessity, then it has an inelastic demand curve—a 1% change in price will lead to a less than 1% change in quantity demanded. The consumer has little choice but to go on buying the good, because there are no other options. Basic foodstuffs are an example of this. If, however, there are lots of alternative goods available, then a 1% change in price will lead to a greater than 1% change in quantity demanded. For example, if bushmeat is just one of many meat products available in a market, then if the price goes up, people will simply switch to an alternative meat.

The elasticity of demand for a good is also important in predicting how increased consumer income will affect demand for a good. For example, in Equatorial Guinea, urban consumers have an income elasticity of 0.26 for bushmeat, 0.55 for fresh fish and −0.27 for frozen fish (their least preferred, but most consumed, protein source, because it is so cheap). This suggests that as incomes rise in the country's oil boom, demand for bushmeat and fresh fish is likely to go up, with potential consequences for sustainability (East *et al.* 2005, Box 6.1).

There's no reason why demand curves for wildlife products should differ from other economic goods—but supply curves are very different. Because the relationship between a population's productivity, in terms of the number of new individual produced, and its size is dome-shaped (Figure 1.1(b)), the quantity supplied only increases with increased price up to the MSY level. After that, although higher prices lead to more effort being put in, that effort yields less and less because the population is depleted (Figure 1.1(c)). Using the simple logistic model, this translates into a backwards-bending supply curve in situations where there is no control over resource use. As Figure 1.2(b) demonstrates, the combination of an inelastic demand curve and a backwards-bending supply curve can potentially be very destabilising (Clark 1990).

1.3.3 Bio-economic systems

Harvesting systems are bio-economic. The population dynamics of the exploited species interact with the incentives that harvester households are experiencing to determine the costs of hunting (both opportunity costs and direct costs; Section 3.2.5.1), and these costs determine the supply curve. When deciding whether,

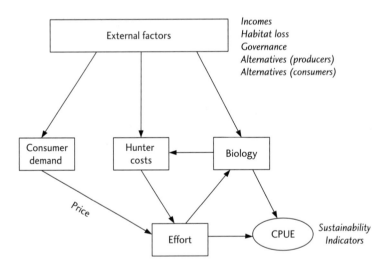

Fig. 1.3 The bioeconomic harvesting system. The relationship between price and hunter cost determines hunter effort, which determines prey population size and structure, which feeds back into hunter costs. This is a dynamic system. Price is a function of consumer demand, and external factors affect the system—demand may be affected by alternative foodstuffs, hunter costs by alternative livelihood activities, prey by habitat destruction and everything by the institutional and governance structures that the system sits within (laws, traditions, etc). An example of a sustainability indicator, catch per unit effort, is given.

where and for how long to harvest, the harvester weighs up the expected costs of each option against the expected benefits (which may be monetary revenues, food for the table, medicines or timber for building). The outcome of these harvester decisions is the harvesting effort expended. The harvest mortality experienced by an exploited population at a particular place and time is proportional to harvesting effort, and this determines biological sustainability. This relationship is shown in Figure 1.3.

One determinant of the stability of a bio-economic system is the institutional structure—and particularly, who has control over harvesting. This determines how backwards-bending the supply curve is, with a continuum between a sole owner who doesn't discount the future (Figure 1.2(a)) and an open access system where people don't consider the future in their harvesting decisions (Figure 1.2(b)). Another determinant is the market structure—if the good has many substitutes, then the demand curve will be nearly flat (Figure 1.2(a)), while if it is an irreplaceable necessity it will be nearly vertical (Figure 1.2(b)), which may lead to big changes in the level of population depletion for small changes in price.

1.4 Book structure

There are five steps towards achieving sustainable use:

- Objective setting—what do we wish to achieve with our intervention?
- Data collection and analysis—what is the current situation?
- Understanding—what factors affect sustainability?
- Intervention—how can we move the system towards the desired state?
- Maintenance—ongoing monitoring and adaptive management.

These steps are covered in the following chapters (Figure 1.4):

In **Chapter 2** we discuss the approaches that we can use to obtain biological data, such as abundance, density and population structure. These data are fundamental

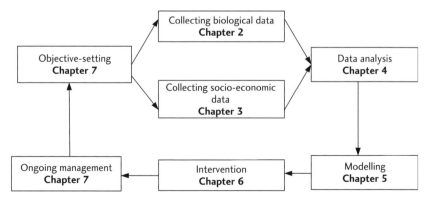

Fig. 1.4 Schematic of the book's contents by chapter.

to an understanding of biological sustainability. In **Chapter 3** we talk through the methods we can use to understand natural resource users' incentives, giving some general hints about collecting socio-economic data, and then going through some examples of studies that take a range of approaches to the human side of sustainability. **Chapter 4** deals with how we obtain indices of current sustainability from statistical analysis of the data collected in Chapters 2 and 3, while **Chapter 5** explains how to use our insights into the dynamics of the system to develop predictive models. These can be used to assess the effects of conservation interventions or external factors on sustainability, and so can be very powerful tools, if the basic understanding and parameter estimation are robust.

In **Chapter 6**, we discuss the range of potential conservation interventions that are being used to improve the sustainability of harvesting systems. These include interventions that are directly targeted at harvesting, but also indirect approaches such as promoting goodwill and cultural value. Finally, in **Chapter 7**, we discuss how to manage systems for long-term sustainability. This includes having a robust monitoring system that gives us the data we need, and having an adaptive approach to management that learns from mistakes and is able to respond to the inevitable shocks and surprises that characterise our world.

It is important to emphasise the inter-relatedness of the issues discussed here, and the importance of having a **framework for understanding** the system and formulating research questions before collecting any data. Otherwise data are collected in a scatter-gun or opportunistic way, wasting time and resources and potentially not answering the questions that are subsequently realised to be important. In an ideal world, data are collected in a cost-effective manner within a robust management system. This suggests that the discussion of project objectives and measures of success in Chapter 7 should be read before embarking on the data collection discussed in Chapters 2 and 3. Chapters 4 and 5, the selection of sustainability indices and modelling of the system, require the understanding and data obtained in Chapters 2 and 3, but how do we know which data we need until we have carried out the analyses found within these chapters? In the real world we often find ourselves facing a crisis situation in which management is non-existent or dysfunctional, we are ignorant about the biology of the species and the threats that it faces, and we need to start somewhere. We have organised the book in the order which we feel will make most progressive sense to the reader, and hope that the messages in the later chapters, particularly concerning the need for clear and costed **objectives** and measures of success, are taken on board before leaping into data collection.

2

Techniques for surveying exploited species

2.1 Scope of the chapter

In Chapter 1, we show how a basic model of population dynamics can help us to understand the principles of sustainability, and in Chapters 4 and 5 we cover the practicalities of a range of methods for assessing and predicting biological sustainability. These concepts and techniques all require some form of biological information on the exploited population. At the simplest end of the spectrum, changes in population size can be used as a direct measure of the state of the system. More complex model-based methods might require the estimation of other parameters such as intrinsic growth rate or age-specific rates of survival and productivity. The range of parameters relevant to exploited species is summarised in Table 2.1. In this chapter, we summarise the basic classes of method available to measure each of these key parameters. The state of the art in most of these techniques requires computer-intensive statistical analysis, but fortunately a large amount of software is available that makes these analyses accessible, much of it free. For detailed instructions on analysis, we therefore refer readers to these programmes and associated literature (Section 2.7). The information provided in this chapter is primarily intended to equip you with the knowledge necessary to make informed choices about appropriate techniques and robust survey designs for your fieldwork.

2.2 Sampling considerations

Just occasionally, it is possible to observe all the individuals in a population, giving a point estimate for population size with no uncertainty attached. With a **complete census** of this kind, we do not accept the possibility that our estimate may be wrong, and make no attempt to estimate how far out it might be. If you're going to use this approach, you therefore need to be absolutely convinced that your observations are complete and unbiased. In practice this is possible only for very conspicuous species in small areas. If we wish to say something about larger populations (as will almost always be the case for exploited species), it is far better to accept that we cannot observe all individuals with certainty and **sample** the population in a way that allows us to make robust inferences about the entire population, based on the observed sample.

However, this approach brings with it the issues of **bias** and **precision**, which should be understood if sampling methods are to be applied appropriately. Ideally,

Table 2.1 Summary of parameters covered in this chapter, their relevance to the assessment of biological sustainability and the sections in which they are addressed.

Parameter	Usual symbol	Relevance to sustainability assessment	Section
Population size	N	Defines the biological state of the system—fundamental parameter of interest.	2.3
Unexploited population size (carrying capacity)	K	Can be used to parameterise logistic growth model, hence define sustainability reference points and predict impacts.	2.3
Intrinsic rate of increase	r_{max}	r at very low population size. Can be used to parameterise logistic growth model, hence define sustainability reference points and predict impacts.	2.4.1
Instantaneous rate of population change	r	Can be used to estimate r_{max} and define population trends.	2.4.1
Finite rate of population change	λ	Alternative definition of r ($\lambda = e^r$).	2.4.1
Survival rate	S	Can be used (with P) to predict rate of population change; can be used to quantify density dependence.	2.4.2
Productivity rate	P	Can be used (with S) to predict rate of population change; can be used to quantify density dependence.	2.4.3
Harvest mortality rate		Gives a direct measure of harvest pressure.	2.4.2
Density dependence		Pattern of change in demographic parameters with density; a fundamental determinant of sustainability.	2.4.4
Rate of body size growth		Can determine value of offtake; can be used to parameterise size-structured models.	2.4.5
Movement rate		Can be used to parameterise spatially explicit models.	2.5.1
Habitat associations		May be necessary to uncover impacts of harvest; can be used to estimate population size over large areas.	2.5.2

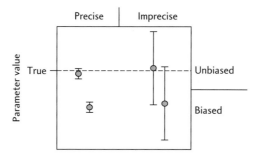

Fig. 2.1 When estimating a parameter such as abundance, there are four possible outcomes, defined by precision and bias that are either high or low. Parameter estimates are indicated by points, with the degree of statistical confidence in each estimate indicated by error bars. The true parameter value is indicated by the dashed line.

we are looking for methods that minimise bias and maximise precision, but of these two factors, low bias is the more important (Figure 2.1). This is because actions based on a biased estimate can be disastrous (e.g. a decision to continue harvest of a depleted population because of a mistakenly high population estimate), whereas low precision should simply increase the caution with which results are treated. We describe below the principles of bias and precision, and outline how to achieve the best possible sample in practice.

Sampling requires the definition and selection of **sampling units**. For example, these may be discrete sites for estimating abundance (Section 2.3), individual organisms for estimating demographic rates (Section 2.4), or individuals, households or communities of people when studying resource users (Chapter 3). The ways in which sampling units are defined and selected have important implications for bias and precision, which we outline in the following sections.

2.2.1 Bias

A key source of bias is the failure to select **representative** sampling units. For example, haphazard selection (throwing quadrats, sticking pins in maps, selecting individuals on encounter, etc.) is not adequate because it allows subconscious selection of sites that 'feel right', or can be influenced by variation in detectability between individual organisms. Either case tends to result in an unrepresentative sample.

A second important source of bias is the failure to meet the **assumptions** of the analysis. Such inappropriate use of a model gives rise to **model error**, which can result in serious bias. Most of the methods described in this chapter not only sample many sites or individuals, but also use some form of model to estimate parameters. It is crucial that the assumptions of these analyses are understood in order to avoid violating them. Specific assumptions are highlighted under each method described in this chapter.

2.2.2 Precision

When sampling a population, values are not estimated with certainty, but with error (the converse of precision). There are two sources of error that influence precision. **Process (sampling) error** is the result of the spatial distribution or other characteristics of the population. Uncertainty arises here because the individual sites or organisms we happen to have selected may by chance fail to be representative of the population, even if the method used is theoretically unbiased. **Observation error**, which may be present either as well as or instead of process error, results from uncertainties in the way in which the population is observed. It is important to recognise the dual sources of error because it can help to clarify the most promising way to improve precision. For example, it may be pointless exerting huge effort to reduce process error by sampling more sites if the real problem is that estimates of detectability at a given site are hopelessly imprecise (i.e. observation error is high).

Lack of precision in estimates is a problem because it can obscure **real differences**, such as significant decline in a population due to harvesting. If one is setting out to detect differences, it is therefore worth first defining the magnitude of difference that you would like to be able to detect, and knowing the degree of precision that will be sufficient to do this. Very good precision (a coefficient of variation of around 3%) is required to detect a change in population size of 10%. Typically, population surveys achieve a coefficient of variation in the region of 10–20%, which only allows the confident detection of a 40–80% change in population size.

In order to decide how much confidence to place in a given estimate (either qualitatively, or when formally testing whether there is a significant difference from some baseline), we need some way to measure precision. The classical **parametric** approach assumes a certain underlying distribution and estimates the parameters of this distribution from the data. For example, given a normal distribution with mean 10 and sampling variance 4, statistical theory allows us to calculate that there is a 95% chance that a value randomly selected from the distribution will lie between 6.08 and 13.92. This gives a confidence interval for the mean that can be compared with other values. This approach has the benefit of an exact statistical formulation, allowing easy calculation, and is the basis of most measures of precision. We therefore provide equations for calculating parameter standard errors throughout this book, and describe relationships between this and other measures of precision in Box 2.1. The major constraint with the parametric approach is that it requires the data to at least approximate the assumed distribution, an assumption that will often not hold. In this case, it may be safer to use **non-parametric bootstrapping** (Box 2.1), which makes no assumptions about the underlying distribution of the data. This is a computer-intensive approach, requiring some ability to write simple programs, although existing software makes the technique relatively accessible these days (see Section 2.7).

Box 2.1 Parametric and non-parametric measures of precision.

By **parametric**, we mean an approach that assumes an underlying probability distribution such as the Normal (Gaussian or bell-shaped) distribution. There are several related parametric measures of precision. The **sampling variance** of parameter $\hat{x}$, denoted $\text{VAR}(\hat{x})$, can be estimated in many ways, depending on the sampling process. In the special case where $\hat{x}$ is a population mean, the sampling variance is a function of the **population variance**, s^2, and sample size, k:

$$\text{VAR}(\hat{x}) = \frac{s^2(\hat{x})}{k}$$

Population variance (often alternatively denoted σ^2) is the standard function available on calculators and spreadsheets, that is, given k individual observations, x_i:

$$s^2(\hat{x}) = \frac{\sum(x_i - \hat{x})^2}{k - 1}$$

The square root of the population variance gives the **standard deviation**, while the square root of the sampling variance gives the **standard error**:

$$\text{SE}(\hat{x}) = \sqrt{\text{VAR}(\hat{x})}$$

The **coefficient of variation** (CV) is the standard error expressed as a proportion (or percentage) of the parameter estimate:

$$\text{CV}(\hat{x}) = \frac{\text{SE}(\hat{x})}{\hat{x}}$$

The parameters discussed in this chapter are generally not normally distributed, however, when sample size is reasonably large, a normal approximation for the **confidence interval** (CI) is often reasonable:

$$\text{CI}(\hat{x}) = \hat{x} \pm t_{2,\alpha}\, \text{SE}(\hat{x})$$

where value $t_{2,\alpha}$ is taken from the two-tailed t-distribution. For a 95% confidence interval, $\alpha = 0.05$, and for large sample sizes, t approaches 1.96. When precision is relatively low, it may be more appropriate to use a log-normal approximation for the confidence interval, giving intervals that are asymmetric and constrained to be positive:

$$\text{CI}(\hat{x}) = (\hat{x}/w,\, \hat{x} \times w)$$

$$w = \exp\left(t_{2,\alpha}\, \sqrt{\ln(1 + \text{CV}(\hat{x})^2)}\right)$$

When there is no good basis for assuming a particular parametric distribution, **non-parametric bootstrapping** can be used. In this approach, the data are allowed to define their own distribution by random resampling. For example, here is a simple bootstrap procedure for calculating the confidence interval for a mean:

- Given k data points (e.g. the numbers of individuals in a sample of plots), pick a random selection of them. The new 'virtual' sample should contain exactly k data points, but sampled with replacement, so that some may appear more than once in the new sample, while others do not appear at all.
- Estimate mean abundance from the new sample, and keep a note of this value.
- Repeat this many times (typically at least 1000 replicates are used).
- The 2.5% and 97.5% percentiles of the resampled abundance estimates then give the 95% confidence interval.

A related non-parametric approach, **randomisation**, can be used to test whether two means are significantly different from one another. Here's one way to do this:

- Randomly re-assign each raw value to one of the two categories, preserving the sample size of each category (a process known as permutation).
- Calculate the difference between category means for the permuted sample.
- Repeat this many times, storing the difference between means each time.
- Count the number of permuted differences that are more extreme than the observed difference (either more negative or more positive than absolute value) and divide by the total number of permutations. This is the approximate probability that the observed difference could have arisen by chance, and is equivalent to the p-value of standard statistical tests.

There are many variations and elaborations of these randomisation approaches. For a simple introduction, including software application, see Howell (2007). A comprehensive text is provided by Manly (1997).

2.2.3 Getting the best possible sample

When designing a survey, it is important to understand the possible sources of bias and poor precision in order to avoid them. Below are some of the key practical considerations.

- Ensure that your sample is representative by either **random sampling**, for example, a study area might be divided into blocks on a map, each block assigned a number and sample blocks selected by picking random numbers using a random number generator, or **systematic sampling**, for example, covering the whole study area with points on a grid system, or with transects running parallel to one another. Care is needed with this approach to ensure that the sampling pattern is not aligned with variation in the population (for example, if transects run along landscape features that also influence abundance). Randomised starting points and/or directions can help to achieve this. The important point here is that locations should not be rejected or avoided simply because they might be hard to access or yield disappointingly few observations.

- Increase precision by increasing the **sample size**. Bear in mind, though, that sampling error decreases with the square root of sample size (Box 2.1), giving diminishing returns on effort.
- Don't artificially improve precision through **pseudoreplication** (treating non-independent sampling units as if they were independent). For example, this might arise if large sampling units are arbitrarily subdivided after the event in order to boost sample size. Occasionally, when surveying sparsely distributed but locally common species, it may be cost effective to increase sampling effort when aggregations are encountered, in which case **adaptive cluster sampling** techniques can be used (Thompson 2002, 2004). Usually, though, it is preferable to design the field study to deliver fully independent replicates.
- Consider using a **stratified sample**. This is the purposeful partitioning of sampling units between areas or groups that differ in ways thought likely to influence the parameter of interest. For example, samples may be stratified by habitat. This has the dual benefit of **improving precision** by accounting for some of the variability between samples (although this only works well when the number of strata is small relative to the total number of samples) and alleviating potential **problems with bias** due to under-sampling of less accessible areas (so long as at least some sampling can take place in these areas).
- Specific **environmental** variables can be measured at each sample site, and a statistical model used to relate spatial variation in the parameter of interest to these variables. As well as helping to increase precision, this approach can provide information about the correlates and potential causes of variation, helping to tease out harvest effects from other potential influences (see Section 2.5.2 and Box 4.4).
- When surveys are primarily aimed at measuring changes over time, precision can be improved by **resampling** the same units on each occasion (Plumptre 2000). However, care must be taken with this approach to ensure that the survey activity itself does not affect the parameter being measured. For example, multiple visits to a plot may alter population size through habitat disturbance.

Comprehensive coverage of sampling theory can be found in Thompson (2002). Online resources providing more details on the subject are suggested in Section 2.7.1.

2.3 Measuring abundance

Abundance is a general term for the number of individuals in a population, which can be measured either as population size or, more usefully, as density of individuals within a given area. Knowing how to proceed in the field depends on knowing how you are going to analyse the data you collect. It is therefore essential to know what analysis methods are available to you, and how they work, so that the data can be gathered in an appropriate fashion, minimising the chances of nasty surprises at the analysis stage. This section provides an introduction to the available statistical methods, followed by some guidance on how to select the most appropriate one in your case.

2.3.1 Plot sampling

Plot sampling refers to any form of fixed area that is searched for the target species. Searched areas may be square quadrants, rectangular strips or belts, round plots or even irregularly shaped regions defined by landscape features if this is most convenient, although extra care needs to be taken to avoid bias in this case. Plot sampling is the most basic analytical method, requiring no estimation of detectability because we assume that all individuals within the plots are found. Given a total of n individuals seen in k plots, each of area a, the estimated density is therefore simply:

$$\hat{D} = \frac{n}{ka}$$

Assuming that individuals are randomly and independently distributed, and that sample plots do not overlap, the population variance, s^2, can be used to measure the precision of the density estimate (see Box 2.1). If the population is in a well-defined area of size A, the population size estimate is simply:

$$\hat{N} = \hat{D}A,$$

$$SE(\hat{N}) = SE(\hat{D})A.$$

However, if there is no overlap between plots (sampling without replacement), this approach increasingly overestimates variance as the proportion of the whole area covered, p, increases. In this case:

$$SE(\hat{N}) = SE(\hat{D})A\sqrt{1-p}$$

so that as proportional coverage approaches 1, the variance approaches 0. This makes sense, because if we observe the entire population, we are no longer sampling, and have complete certainty in the population count. The assumption that individuals are independently and randomly distributed, required by the above variance estimators, will frequently be violated in practice. More often, organisms will be more clumped than expected, leading to under-estimation of variance under the above approximations. In this case **non-parametric bootstrapping** may be more appropriate (see Box 2.1).

The requirement that all individuals in a plot should be found generally restricts plot sampling to **conspicuous species** that don't hide or flee from people. This typically means sessile species, particularly plants; however, the method can occasionally be applied to more mobile species (e.g. primates in small forest patches; Mbora and Meikle 2004; Anderson *et al.* 2007; Box 2.2). For territorial animals, repeat visits over time can be used to **map territories**, and so arrive at a total count of territorial individuals (Bibby *et al.* 1992; Fuller *et al.* 2001). However, territory mapping is very time consuming, and relies on territorial signals such as calls.

Box 2.2 Plot sampling in action: underwater visual census for fish.

Underwater visual census (UVC) is a standard technique used to measure the abundance of fish in clear, shallow water, e.g. over coral reefs (English *et al.* 1997; Figure 2.2). Ashworth and Ormond (2005) used this method to assess the efficacy of a marine protected area (MPA) in the Egyptian Red Sea. Transects of 100 m were swum by divers, counting and identifying all fish within 5 m either side, giving sampling plots of 1000 m². This is somewhat larger than the plots generally used in other UVC studies, a deliberate strategy to reduce the variability in counts between plots, and so increase the precision of estimates. However, this strategy carries the risk of missing significant numbers of fish, so underestimating density, and is therefore justified only where there is exceptional visibility. Because of the diversity of species covered, it was not possible to complete each transect in one pass. Transects were therefore swum several times, focussing each time on one or a few related fish families. Transects were placed inside and outside the MPA, and at three different depths. Twelve transects were placed in each depth/location combination, and the mean fish density (per m²) for each combination is therefore the total number of fish seen divided by 12,000. At the shallowest depth, seven fish families had significantly lower abundance outside the MPA, but this difference was not seen in deeper water, reflecting lower fishing pressure at this depth.

Fig. 2.2 Divers in northern Mozambique counting fish in strips of fixed width in order to estimate abundance. Training in distance estimation is required in order to be able to judge whether fish seen are within the strip. The method only works in very clear water. Photos: © Cabo Delgado Biodiversity and Tourism Project.

UVC is a somewhat unusual application of plot sampling in that it is applied to species that may be highly mobile. This can lead to heavily biased estimates, either under-estimation caused by disturbance, or in some cases over-estimation caused by attraction to the surveyor's activity (Edgar *et al.* 2004). In principle, more sedentary fish species would be ideally suited to plot sampling, but these species are often difficult to see, and it would probably be impossible to search the large plots used in this example with sufficient confidence. Smaller plots searched more carefully would be appropriate in this case, but even then densities of species that hide themselves in the substrate may be underestimated hugely (Willis 2001).

It also misses the non-territorial sector of the population, which can be significant, both ecologically and in terms of harvest offtake. For example, Martin (1991) found that when male willow ptarmigan *Lagopus lagopus* were removed from their territories, 70% of them were quickly replaced by unpaired males.

2.3.2 Nearest neighbour (plotless) sampling

This method is aimed at **static species**, particularly plants. It works by randomly selecting a number of individuals or points within the study area, and measuring the distance to the nearest neighbouring individual. This is actually conceptually similar to plot sampling, but with a fixed number of objects per plot and variable plot size rather than *vice versa*. Given a sample of n distances, d_i, in a study site of area A, the population estimate is:

$$\hat{N} = \frac{nA}{\pi \sum_{i=1}^{n} d_i}$$

Confidence intervals for the estimate can be calculated using the non-parametric bootstrapping procedure described in the previous section. For further details see Borchers *et al.* (2002).

While this technique is appealingly simple, it has a major flaw, in that it gives seriously biased estimates if the study subjects are not randomly and independently distributed. Because this is rarely true in practice, plotless sampling is very rarely a reliable method. Trying to locate nearest neighbours can also be surprisingly time consuming.

2.3.3 Distance sampling

When searching for any object, the further it is from you, the less likely you are to see it. Distance sampling works by using survey data to quantify this decline in the **probability of detection**, which then allows you to infer how many objects were there but not seen. It works like this. While searching for your target species, instead of pre-defining a fixed area within which you assume everything will be found (as in plot sampling), count all individuals seen, regardless of where they are, as well as measuring how far away each is. This can be implemented either as line transects (equivalent to strip plots) or as point counts (equivalent to circular plots). Line transects require that perpendicular distances between object and transect line are recorded, while point counts record the distance from observer to object.

When the survey is complete, plot the frequency of observations against distance and fit a curve to the data (Figure 2.3). This curve, known as the **detection function**, allows you to calculate an estimate of the proportion of objects present that were seen. Conceptually, the area under the detection function represents the total number seen, while the area within the rectangle bounded by the maximum distance and the maximum of the detection function represents an estimate of the total number present. The ratio of these two areas therefore gives the proportion seen, p. Given a certain number of objects seen, n, density can then be simply estimated by using p as a correction factor, and dividing by the effective area surveyed. For line transects of total length L and maximum detection distance w, density is given by:

$$\hat{D} = \frac{N}{Lwp}$$

Alternatively, for a total of C point counts, density is given by:

$$\hat{D} = \frac{N}{C\pi w^2 p}$$

Fitting detection function models to data is made easy and accessible by free software DISTANCE (Section 2.7.1). Literature associated with this programme (Buckland *et al.* 2004 and programme help files) covers in detail the analytical

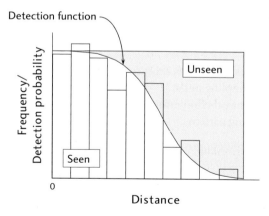

Fig. 2.3 A hypothetical detection function fitted to line transect data, indicating seen and unseen portions of the population. The transect line is at zero distance.

philosophy and techniques for this method, and is essential reading for those wishing to analyse distance data. Here we summarise the basic underlying concepts that must be understood in order to design and implement an effective distance sampling survey (Box 2.3).

Precision in distance-based estimation depends on having both a reasonable number of replicates (lines or points) to minimise sampling error, and a reasonable number of sightings, to minimise error in the detection model. Although it is possible to estimate sampling error without replicates by assuming a theoretical sampling distribution, this approach is not ideal because the assumed distribution is likely to be wrong, and it provides a very weak basis for inferring anything about the wider population. As a rule of thumb, at least 10 independent lines or points are needed to get a reasonably precise estimate of the encounter rate. The number of sightings, n, is approximately related to precision by:

$$n \approx \frac{3}{\mathrm{CV}(\hat{D})^2}$$

with coefficient of variation, $\mathrm{CV}(\hat{D})$, expressed as a proportion (Buckland *et al.* 2004). Thus to achieve a coefficient of variation of 25%, around 50 sightings will be needed. Considering the relatively low power to detect differences at this level of precision (Section 2.2.2), this number of sightings can be regarded as a rule of thumb minimum. Where it is difficult and time consuming to set up transects, and encounter rates are low, it might be sensible to make **repeat visits** to transects in order to boost the number of sightings. However, visits to a single transects should not be treated as independent replicates for the purposes of calculating sampling error (see Section 2.2.3 on pseudoreplication). As far as possible, any additional effort should be channelled into new transects rather than repeat visits to existing transects.

When **placing transects**, the ideal randomised or systematic approaches (Section 2.2.3) may be impossible to achieve in practice. It may be that some parts of the study site are simply too hard to get to, or that travelling anywhere in the site is too difficult to place samples far from access points or from one another. For transects, the costs of travel might be minimised by:

- Running long transects with short sections 'off effort', thus dividing the long transect into sampling units;
- Eliminating time off effort entirely by defining sampling units as the arms of a zigzag travelling pattern.

While neither of these approaches provides strictly independent sampling units, it is usually reasonable to treat them as if they are, so long as coverage of the entire study site is reasonably high. Another approach sometimes used in difficult habitat is simply to take a **path of least resistance**, avoiding dense cover or even using paths or roads. This approach is almost certain to yield biased abundance estimates, and should be avoided.

Valid density estimates using distance sampling depend on a number of **assumptions**:

- Detection at zero distance is certain.
- Objects are detected at their initial location.
- Distances are measured accurately.
- Sightings are independent of one another.

These assumptions can be relaxed under some circumstances, but this generally requires additional information, and it is preferable to design surveys to satisfy the assumptions whenever possible. It is therefore very important that they are understood at all stages of a survey, from planning, through implementation to analysis.

Detection probability at zero distance

The assumption of certain detection at zero distance is the most important, and whenever possible, the study design should ensure that it is satisfied. This is best achieved through **training and motivation** of surveyors to ensure that the assumption is understood, and that the central line or point is searched thoroughly. However, overenthusiastic searching at the centre, at the expense of effort further out, can lead to a very rapid decline in numbers seen with distance. This is problematic at the analysis stage, since models fitted to such 'spiked' data are highly sensitive to minor fluctuations in frequencies, and give poor precision. Ideally, the detection function should have a 'shoulder', being reasonably flat for some distance from the centre before declining (Figure 2.4). This requires a balanced approach to searching, ensuring that the centre is well covered, but not neglecting greater distances.

In cases where detection at the centre is unavoidably **uncertain**, a density estimate can still be achieved by making an independent estimate of detection probability at the centre. The most common way to do this is by deploying two or more independent observers on each line and recording the incidence of cases not seen by all observers (Borchers *et al.* 1998). For this to work, it is essential that the observers neither disturb the target objects before the others have a chance to observe them, nor give cues to the others when they find an object.

Detection at initial locations

The assumption that objects are detected at their initial location is often a problem for mobile animals. Movement that is **not influenced by the observer** is not a problem if the speed of movement is less than about half that of the observers. However, if movement is much faster, care should be taken to ensure that, as far as possible, a snapshot of animal distribution is taken within a short space of time. Failing to do this risks animals effectively accumulating in the counted area as they pass through, leading to over-estimation of density. This is a particular problem for point counts of highly mobile animals.

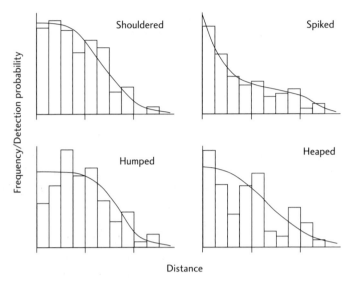

Fig. 2.4 Examples of line transect data with fitted detection functions. The ideal form has a broad shoulder of high detectability close to the line (shouldered). The other forms commonly observed in data make robust analysis difficult, and should be avoided through appropriate field methods as far as possible.
Spiked data usually arise from excessive attention to the centre line or point.
Humped data are usually the result of animal movement away from the observer.
Heaped data are usually the result of approximating distances to the nearest round number.

More problematic are cases where animals **respond to observers** by moving away from or, less commonly, towards them (Box 2.3). When such movement is either ignored or not detected, it effectively results in an anomalous density around the observer. In the case of animals fleeing, this leads to under-estimation of density. In practice, it will usually become quickly obvious to observers whether this is a problem, although occasionally movement may take place in thick cover and remain undetected. In this case, the problem might show up only as a humped detection function (Figure 2.4). This problem cannot be fixed at the analysis stage, and every effort should therefore be made to ensure that distances are measured to the animal's initial location, before any movement in response to observers has taken place. Tricks for helping to ensure this might include:

- Using multiple observers to define the extent of responsive movements, keep track of animal locations and increase the chances of early detection (although this can be counterproductive if the additional observers increase disturbance);
- On line transects, looking well ahead to detect responsive movements as early as possible;

- On line transects, instead of attempting to measure the perpendicular distance on the ground, measuring the direct distance d, together with the angle θ between the transect and the line of sight to the animal. Perpendicular distance can then be calculated later using trigonometry: $d_{\text{perp}} = d_{\text{direct}} \cdot \sin \theta$. This allows the animal's location relative to the transect to be pinpointed rapidly, minimising the difficulties of trying to keep track of moving animals while attempting to move into position level with them on the transect.

Accuracy of measurements

Given a rigorous approach to measurement and the appropriate tools, this assumption is easy to satisfy. Whenever possible, accurate measurements should be made with the aid of tapes, range finders, graduated markers attached to observation platforms, clinometers or sighting compasses, as appropriate to the environment, the form of transport and the target species. However, where limited resources mean that suitable equipment is unavailable, it may be possible to **estimate distances** by eye. As well as being cheap, this approach has the benefit of being quick and easy to apply in the field. However, you should be alert to two serious problems with estimation by eye:

- There is likely to be consistent bias in distance estimates, leading to biased density estimates (if distances are underestimated, densities will be overestimated).
- Estimates are liable to be approximated to the nearest round figure. This leads to heaping of the distance frequency data around the more popular numbers (Figure 2.4), giving a poorer fit of the detection function and reducing precision.

If distances must be estimated by eye, it is essential to spend time training observers to do this accurately in relevant habitat and weather conditions. Observers should be aware of the need to avoid rounding distances, and surveys should not proceed until all are able to make unbiased estimates of distance.

In some cases, it might be easier to assign sightings to **distance categories** in the field, rather than recording full distance information. This is acceptable, although accurate distance measurements are still important in this approach; assignment of sightings to the wrong categories has the same effect as biased recording of distances. It is also necessary to use a reasonable number of distance categories—an absolute minimum of three categories is required to make meaningful modelling of the detection function possible.

Independence of observations

For species that form close-knit **groups** (particularly social animals), individuals are not observed independently of one another. This can be ignored without giving rise to biased density estimates, but the non-independence of individuals will give rise to artificially narrow confidence intervals. Realistic estimation of error therefore demands that the unit of observation is the group. In this case, distance analysis provides an estimate of the density of groups, and this is multiplied by the mean group size to give an estimate of overall density, with overall variance incorporating a component of variance in group size. In the field, perpendicular distances to group centres must be measured, and this should be done directly during the survey rather than measuring distances to nearest individuals and correcting for group spread later. Such adjustment of distances after the event introduces further error, and can lead to bias if group spread varies over time and space (Buckland *et al.* 2004).

Although group counts should ideally be made during the distance survey, this can lead to **biased group size estimation**. For example, larger groups may be more conspicuous, leading to over-estimation, or alternatively some individuals in more distant groups may be missed, leading to under-estimation of group size. There are three possible ways around this problem, at either the data collection stage, the data selection stage or the data modelling stage. During data collection, efforts can be made to ensure that all group counts are accurate, at least up to a certain distance away. To do this, it is acceptable to leave the line or point if necessary to get a more complete count, although any new groups seen while away from the line should be ignored for the purposes of abundance estimation. At the data selection stage, if it seems clear that group counts were good up to a certain distance, data beyond this distance can be discarded from the analysis (termed data truncation). Finally, if neither of these approaches works well, it is possible to model variation in apparent group size with distance, thereby correcting for incomplete data to give an unbiased estimate of group size (Buckland *et al.* 2004).

Box 2.3 Distance sampling in action: line transects for duikers.

Duikers are small African forest antelopes which are very commonly hunted. They are usually very shy, and found in dense vegetation, making them extremely difficult to see. Newing (1994) tested a range of methods for estimating the density of several species of duiker, primarily Maxwell's duiker *Philantomba maxwelli*, in and around Taï National Park, Côte d'Ivoire, including distance sampled line transects. In secondary vegetation, pilot transects yielded no sightings of duikers due to extremely dense vegetation, so distance sampling had to be abandoned there. However, in closed canopy forest with a more open shrub layer, sightings were frequent enough to allow density estimation. Moving as quietly as possible at around 1 km/h, transects were carried out both by day (yielding 41 sightings from 33 km of transects) and by night using torch-light (yielding 46 sightings from 23 km of transects).

Distance analysis of daytime data suggested evasive behaviour (Figure 2.5), which was confirmed by radio-tracking data, and by occasional observations by other researchers of duikers fleeing surveyors who were entirely unaware of the duiker's presence! In contrast, there was no evidence of evasion during the night, when duikers tend to freeze in the torch beam rather than fleeing. In an attempt

Fig. 2.5 Duikers such as this blue duiker, caught in a snare in Equatorial Guinea, are an extremely common target for bushmeat hunters in west and central Africa. Unfortunately, it's very difficult to estimate their abundance using distance sampling because they are good at moving away from people before being detected, as shown by the humped day time detection function. The result is better using torchlight by night, but this is much harder work for the surveyors. Photo © Noelle Kumpel.

to get round the problem of evasive movements during the day, all observations within 4 m of the transect were discarded prior to analysis (a technique known as left-truncation). However, despite left-truncation, duiker density estimated at night was around three times higher, and presumably more accurate, than the daytime estimate (estimates and 95% confidence intervals, daytime: 36 km^{-2} [24–56]; night-time: 101 km^{-2} [68–150]). As is usually the case, left-truncation was unable to fix the problem of evasion.

2.3.4 Mark-recapture

This technique is generally applied to animals that are difficult to see in the wild, but which can somehow be recognised as having previously been observed when encountered. Usually this is achieved by catching, marking and recapturing individuals. By collecting information on the individuals that are captured and recaptured over two or more different occasions, it is possible to estimate the proportion of the population that has been captured, and so calculate the total population size (Box 2.4). In essence, the method works as follows. On the first of two capture occasions, individuals are caught and marked in some way, then released to mix with the remaining individuals that were not caught. On the second occasion, provided that thorough mixing of marked and unmarked individuals has taken place, the proportion of individuals in the sample that are marked provides an estimate of the proportion of the population that was originally marked. The number in the population is then simply the number caught on the first occasion divided by the proportion of individuals in the second sample that bear a mark. This common-sense calculation is in fact biased at realistic sample sizes, and although this bias can be corrected (giving rise to the well-known Lincoln-Peterson method; Seber 1982), in practice this approach is seldom used because more complex methods are required to detect and correct for various other sources of bias.

Modelling of mark-recapture data is implemented in a variety of specialist software (Section 2.7.1). Full treatments of mark-recapture methods for estimating abundance can be found in Seber (1982, 1986, 1992), Pollock *et al.* (1990) and Williams *et al.* (2002). Here we outline the key considerations for setting up a robust survey.

Precision in mark-recapture estimates of abundance depends principally on the **proportion of the population** that is captured. As the proportion captured decreases, the confidence interval widens rapidly, approaching infinity as the proportion captured approaches zero. A substantial proportion of the population should therefore be captured, preferably at least 50%. Unfortunately, because we do not initially know how large the population is, or how rapidly individuals can be caught, this does not immediately help to determine how much effort will be needed for a mark-recapture survey. However, an assessment might be made based on a conservative guess at the minimum expected density and probability of capture. Alternatively, the data might be monitored over an indefinite number of capture occasions in order to assess when a sufficient proportion of the population has been caught. In either case, it is usually necessary to have more than two capture occasions in order to achieve a reasonable overall capture rate.

If the study area is relatively small, self-contained and easily defined (e.g. an enclosure or habitat island), capture effort can be spread throughout to provide an estimate of the total population size. Because the site has well-defined boundaries, the area is also known precisely, allowing density to be calculated easily. More usually, capture effort must be concentrated in a small part of the wider study site, in which case the **effectively sampled area** is not clear. A minimum sampled area can be defined by a line drawn around the outer-most capture points, but because individuals from outside this region are also susceptible to capture, the effectively sampled area is greater than this. This additional area can be defined as a boundary strip around the minimum sampled area, the width of which is defined by the target species' typical patterns of movement. Boundary strip width can be measured in one of two ways:

- Based on the differences in density between nested sub-grids in a regular square grid of traps (Seber 1982; Jett and Nichols 1987). This method requires a large number of traps and a large sample size to give a reliable estimate (Wilson and Anderson 1985a).
- As half the maximum distance moved between capture locations, averaged across individuals (Wilson and Anderson 1985b). This method is unreliable if individuals are rarely recaptured more than once or twice.

When the area sampled by a mark-recapture survey does not cover an entire population, it is important to realise that an isolated mark-recapture survey is effectively only a single replicate, and does not therefore provide a sound basis for inferring anything about the wider population. If the intention is to estimate abundance over a wider area, a set of replicate mark-recapture surveys should be carried out, selecting survey locations according to the criteria set out in Section 2.1 in order to ensure a representative sample.

The basic mark-recapture method described above requires a number of restrictive **assumptions** to be satisfied if it is to give unbiased estimates of abundance:

- All individuals are equally likely to be caught;
- The population remains constant between capture occasions (i.e. it is closed);
- There is no loss or misidentification of marks.

It is rare that all of these assumptions will be satisfied, and foolhardy to assume so without good evidence, as uncorrected variation in capture probability leads to biased abundance estimates. Fortunately there are many techniques implemented in the available software that can allow violations of assumptions to be detected and controlled for. In this section, we give an overview of the most common reasons for violations, and the ways in which they might be avoided through good survey design and appropriate analysis. The analyses introduced below generally require complete capture histories for all individuals to be known, and individual-specific marks should therefore be used whenever possible.

2.3.4.1 Equal catchability

Common reasons for variation in capture probability are poor coverage of the surveyed region, intrinsic variation between individuals in their susceptibility to capture, changes in overall capture probability with time and changes in individual behaviour as a result of capture. **Poor coverage** can arise if capture effort is too thinly spread, leaving holes in the effectively sampled area in which some individuals may have no chance of being caught (Karanth and Nichols 1998). This problem can be avoided by spreading capture effort evenly across the sampled region and ensuring that it is sufficiently concentrated. When captures are made by traps at fixed locations, aim to place at least two traps per minimum home range area of the target species. As a rule of thumb, this approximately translates to a spacing of half the diameter of a home range.

Poor coverage can be one source of heterogeneity in capture probability between individuals, which is a major problem for mark-recapture estimation (Box 2.4). Heterogeneity may also be due to **intrinsic differences** between individuals, for example, if larger individuals are easier to catch. If heterogeneity exists but remains uncontrolled for, abundance will be underestimated, often substantially so. This problem is minimised if a very large proportion of the population is caught, but must otherwise be tackled at the data analysis stage. One way to do this is to seek measurable individual characteristics that might correlate with capture probability (Box 2.4). These characteristics, known as **covariates** in data analysis, might include age, sex, size or location caught. Potential covariates of this kind should be recorded in the field whenever possible. However, even after controlling for observable covariates, unexplained differences in catchability between individuals often remain. To deal with this, methods are available that model individual heterogeneity in capture probability without the need for covariates, but these cannot be relied upon to deliver unbiased estimates if data are sparse.

Changes in capture probability over time are most likely to arise as a result of variation in capture effort, although it is also possible that changes in catchability occur across the population during the survey. If such variation exists without being controlled for, abundance will be overestimated. It is therefore important to test whether significant variation exists by modelling changes in capture probability over time. However, this reduces precision, and it is therefore best to standardise effort to reduce the chance of variation if possible.

Animals may respond to capture, either by becoming more wary, and therefore less liable to capture in future, or, if baiting is used, by seeking out the bait, thus increasing their chances of capture. Where traps are used, these **behavioural responses** are known respectively as trap shyness and trap happiness. Trap shyness might be minimised by using the least invasive form of capture available, while trap happiness might be reduced by setting out bait for a period prior to capture in an attempt to get all individuals to the trap happy stage. If such field methods are inappropriate or ineffective, an analytical model can be used that controls for behavioural responses, although the greater model complexity again reduces precision.

2.3.4.2 Population closure

The assumption of population closure is violated if, during the survey, individuals enter the population through **immigration or birth**, or leave it through **emigration or death**. If the rate of capture is fast enough to accumulate an adequate sample within a short space of time relative to the rate of population turnover, the assumption will be fully or approximately satisfied. However, if the survey takes place during a period of rapid change due to migration, or is so extended that significant numbers of births or deaths are likely to take place, it is necessary to estimate rates of flow into or out of the population in order to control for them and so provide unbiased abundance estimates. This can be done using the same form of data, but using open population models that control for turnover (Sections 2.4.2.2 and 2.4.3.2).

An important consideration in relation to population closure is the impact that capture and marking may have on demographic rates. An example of such an effect is the clipping of multiple toes in amphibians in order to identify them on recapture, which has been shown to substantially reduce survival rates in some species (McCarthy and Parris 2004). As well as being ethically questionable, such impacts of research on the study species are self-defeating, because they prevent objective inference about the population in its natural state. Every possible effort must therefore be made to ensure that capture and marking do not impact on the study subjects.

2.3.4.3 Permanence of marks

When individuals are identified by external tags or marks, it is possible that these may be lost. This is most likely to be a problem in studies focussing on survival rates, in which longer intervals between captures are used, but may still be a problem when estimating the abundance of more delicate species that cannot easily be permanently marked. Where natural markers such as genetic or coat patterns are used, misidentified individuals can have a similar effect, which is to underestimate recapture rate, and therefore overestimate abundance.

Fortunately, if the rate of tag loss or misidentification is known, it can be corrected for at the analysis stage. If there is any suspicion that there might be significant loss of marks during the survey, it is therefore important to measure the frequency with which it occurs. This can be done by double-marking at least a sample of the individuals caught, and recording the proportion of recaptures that retain only one of the marks. Various models are available for estimating rates of tag loss from patterns of double-tag retention (Bradshaw *et al.* 2000; Rivalan *et al.* 2005).

2.3.4.4 Data requirements for mark-recapture

While simple mark-recapture methods can be applied using just a single recapture occasion without identification of individuals, in practice this approach is almost certain to yield strongly biased abundance estimates because of violations of the assumptions. In order to guard against this, more complex modelling is required, which requires more extensive data. This should usually be derived from several capture occasions, with some means of recognising individuals in order to yield complete capture histories for all individuals. In cases where individual-specific

---

Box 2.4 Closed population mark-recapture in action: tagged crayfish and paint-balled elk.

Mark-recapture methods are most often applied to small or medium-sized animals that can easily be caught in large numbers and individually marked. Jones *et al.* (2005) provide a good example of this, applied to a sustainability assessment of crayfish *Astacoides granulimanus* harvesting from forest streams in Madagascar. Streams were searched in 100 m sections, and all crayfish larger than 22 mm were marked with an individually recognisable visible implant elastomer tag. A total of 74 sections were covered, across a range of harvesting intensities, and each was visited five times on consecutive days. This was a sufficient number of visits to provide the information necessary to estimate capture probability, with gaps between visits long enough to allow the population to mix, but short enough to justify the assumption of a closed population (i.e. no births, deaths or migration). In total, 26,096 crayfish were captured 44,286 times. In the analysis, a range

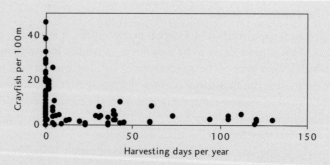

Fig. 2.6 Even relatively light harvesting pressure can dramatically reduce the abundance of crayfish in forest streams in Madagascar. These results were derived from intensive mark-recapture estimates of abundance.

Source: Jones et al. (2005).

of models were tested, including the possibilities of size-dependent capture probability, and of different first and subsequent capture probabilities. Rather than selecting a single 'best' model, capture probabilities and population sizes were estimated by averaging across all candidate models, weighting model contributions by the strength of evidence in the data (program MARK includes a model averaging facility to do this). The results indicated no discernible relationship between harvesting pressure and abundance for smaller crayfish, but a very strong reduction in the abundance of large crayfish in sites that were harvested to any significant extent (Figure 2.6).

For larger animals, practical difficulties and risks to the animals usually prohibit capture on a sufficiently large scale to make mark-recapture a viable option. However, Skalski *et al.* (2005a) got round this problem by marking elk *Cervus elaphus* with paint-ball guns fired from helicopters in a remote part of Washington State, USA. In a single day, 55 elk were marked in a discrete population defined by watersheds. During a two hour flight four days later, 54 elk were seen, of which 36 were marked. This provides the necessary information for a simple two-sample Lincoln-Petersen estimate of 82 elk in the population. However, this result is likely to be an underestimate caused by violated model assumptions. At the time of the survey (late March), elk form herds which would have been quite stable over the short period between marking and resighting, and there was probably therefore too little mixing of marked and unmarked individuals. Added to this, larger herds tend to be easier to find, resulting in individual heterogeneity in resighting probability. Because the marking method does not allow recognition of individuals, it is impossible to detect and correct for this heterogeneity, so it is essential to try to avoid the problem through careful survey design. In this case, Skalski *et al.* (2005a) recommend carrying out surveys in the summer, when herds are more fluid, smaller and less variable in size. This runs counter to previous practice favouring winter surveys, when large herds allow many individuals to be marked in a short space of time.

markers are not practical, marks specific to the capture occasion can be used, and numbers of individuals having each possible capture history reconstructed later, but this approach is not appropriate if individual covariates (such as size or age) need to be included in the analysis.

2.3.5 Offtake-based methods

In principle, offtake (which is synonymous with catch or harvest) can be used as an index of abundance. For example, harvest records alone have formed the basis for some key analyses of exploited species' dynamics, most notably Canadian lynx *Lynx canadensis* (Elton and Nicholson 1942) and red grouse *Lagopus lagopus* (Hudson 1992; Haydon *et al.* 2002). These studies used long time series from populations undergoing extreme fluctuations in order to look for retrospective patterns in the data, and the sensitivity of catch as an index of abundance was therefore not a major issue. In general, catch on its own is not a good indicator of the current state of a population, for reasons discussed in Chapter 4. However, given some additional information, catch data can be used to estimate abundance. This is an attractive prospect for exploited species monitoring and research because catch data are usually readily available, avoiding the need for further sampling effort. Unfortunately, these methods are particularly prone to bias. Nevertheless, they may be the only methods available in some cases, and they remain a key tool in fisheries management.

The main categories of catch-based abundance estimation are catch–effort methods (based on monitoring offtake, usually in combination with harvesting effort), the change in ratios method (based on monitoring the offtake of different categories of organism), and catch-at-age methods (based on monitoring the age structure of harvest). All of these methods require **substantial harvesting pressure** so that changes in catch reflect changes in abundance, rather than being overwhelmed by sampling error. The methods are therefore only appropriate for heavily exploited species.

2.3.5.1 Catch–effort methods

These methods minimally require data on the size of a catch, alongside some form of population abundance index measured before and after harvest. To grasp the basic principle, suppose that 100 individuals are harvested, and that the abundance indices before and after harvest are respectively 1 and 0.6. This would suggest that the harvest reduced the population by 40%, so the original population size can be estimated as 100/0.4 = 250. Usually (but not always) **Catch Per Unit Effort** (CPUE) is used as the abundance index, and a series of catches and associated efforts are recorded over time. Box 2.5 gives the details of how to apply this method in practice. Catch–effort estimation of abundance makes a number of important **assumptions**:

- All individuals are **equally likely** to be caught or detected;
- The population is **closed** (apart from offtake);

- The catch is measured accurately;
- CPUE is **directly proportional** to abundance.

Equal catchability of individuals is often a problem, for the same reasons as in mark-recapture analysis. The difference here is that the data contain no information on heterogeneity in catchability, so we cannot estimate and control for it. This is a major disincentive to the use of the method, or, at least, a reason to use it with great care.

Again similar to mark-recapture methods, the **population closure** assumption can be met by carrying out data collection over a short time scale relative to the natural rate of turnover in the population. However, this might be more difficult in the case of offtake-based methods if the removal itself stimulates immigration by creating unoccupied space attractive to colonising individuals, or if the harvesting activity triggers emigration. This problem cannot be detected or controlled for using catch data alone, and results should be regarded as suspect if there is any suspicion that migration might be occurring. Even a relatively low rate of natural mortality or emigration during the process can lead to dramatic under-estimation of abundance, while net natural growth in the population can lead to equally dramatic overestimation. Alternatively, it is possible to avoid the need for the assumption by using an **open population model** instead of the closed model shown in Box 2.5. In this case, catch and effort data collected annually can be used to estimate key population parameters such as intrinsic growth rate, as well as population size. This approach is covered further in Section 4.3.3.

If removal methods are applied experimentally, the assumption of **accurate measurement of catch** is usually trivial (assuming that you keep good records), but

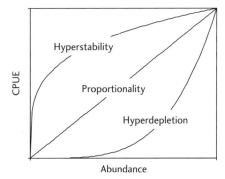

Fig. 2.7 Possible relationships between catch per unit effort (CPUE) and abundance. The ideal is direct proportionality, as this allows standard catch–effort methods to be applied. However, CPUE may remain relatively high across a wide range of population densities, only declining significantly as the population approaches extinction (hyperstability). Conversely, CPUE may decline strongly at high abundance, and remain very low over most of the abundance range (hyperdepletion). Note that other relationships may exist.

if many harvesters must be monitored in order to estimate the total harvest, it can be very difficult to satisfy this assumption. Approaches and pitfalls in measuring catch from harvesters are discussed in Section 2.4.2.4.

The crucial **assumption** behind catch–effort analysis is that the abundance index used is **directly proportional to true population abundance**. Unfortunately, this assumption is often likely to be violated, particularly when CPUE is used as the index. Of the two types of violation (hyperdepletion and hyperstability, Figure 2.7), hyperstability is both the most likely to occur (Harley *et al.* 2001), and the outcome with more serious consequences for conservation because it leads to overestimation of abundance. While catch–effort data alone cannot tell us whether the proportionality assumption is violated, some idea can be gained by assessing whether any likely causes of violation are present in a given case (Table 2.2).

Of the issues raised in Table 2.2, **inappropriate analysis** is clearly the one that can most easily be avoided. This requires that data are gathered at a scale at which sampling is essentially homogeneous, ensuring that no significant areas are left

Table 2.2 A summary of the key reasons why the assumption of proportionality between catch per unit effort and abundance might be violated.

Type of violation	Type of cause	Specific cause
Hyperstability	Animal behaviour	Groups remain easy to find, despite reduced group sizes
	Animal behaviour	Individuals aggregate more as depletion proceeds due to habitat selection or conspecific attraction
	Harvester behaviour	Offtake limited principally by high handling time rather than searching
	Harvester behaviour	Increasing efficiency of harvest method over time
	Inappropriate analysis	Aggregating data over a wide area when harvesters continually move to patches with the highest abundance
Hyperdepletion	Animal behaviour	Individuals learn to avoid capture
	Animal behaviour	Heterogeneous capture probabilities— vulnerable individuals are caught first
	Harvester behaviour	Interference between harvesters
	Harvester behaviour	Decreasing efficiency of harvest method over time (see Box 2.5)
	Inappropriate analysis	Aggregating data over a wide area when harvesters remain in heavily depleted localities while the wider area remains less depleted

Note: Hyperstability and hyperdepletion are defined in Figure 2.7.

without harvesting effort. If abundance is to be calculated over large, heterogeneous areas, an aggregated CPUE should be calculated as the mean of local values weighted by area, not total catch over total effort (Walters 2003). This clearly requires at least some harvesting effort in all parts of the area; no inference can be drawn about any areas lacking significant harvesting effort. The **spatial scale** of sampling must therefore be chosen on the basis of the timescale of sampling and patterns of harvester movement within that timescale. The problem of **changing harvest efficiency** might be avoided by restricting analysis to periods of constant harvesting technology and behaviour. However, problems arising as a result of either **harvester or animal behaviour** often prevent the reliable use of catch–effort methods. For example, *Pteropus* fruit bats are commonly shot at colonial roosts on Pacific islands (Brooke 2001; Brooke and Tschapka 2002), and because these roosts can be large and stable, many bats might be harvested before any reduction in CPUE can be detected. Caution is therefore needed unless a strong case can be made for the absence of such behaviour-related problems.

Box 2.5 Estimating abundance from catch data, and the method in action: pig harvest in a National Park.

Given two consecutive abundance indices, a_1 and a_2, and catch C, initial population size is:

$$\hat{N}_1 = \frac{C}{1 - a_2/a_1}$$

There are two possible ways to apply this model:

- An independent abundance index might be used, for example, a sighting rate from direct counts of live animals, or the density of fresh dung (but see Section 2.3.6.1 for caveats on indirect sign indices).
- The variation in harvesting effort might be quantified, for example, total time spent searching, total distance travelled by harvesters, or the total number of traps or fishing lines set. In this case, CPUE is used as the abundance index. In this approach, both catch and effort are usually quantified in full, but this does not have to be the case; while the absolute total catch is required, CPUE may be estimated from a representative sample of the harvest if this is more feasible.

In practice, results are much more robust when based on a series of several harvesting periods over which the population becomes progressively depleted. This approach requires a model of population size from each period to the next (a **dynamic model**). A very simple model for this purpose defines fitted population size at one point in time as the previous fitted value minus the observed catch:

$$\hat{N}_{t+1} = \hat{N}_t - C_t$$

A fitted catch value is then calculated for each period as a function of the fitted population size and observed catch and abundance index:

$$\hat{C}_t = q\hat{N}_t C_t / a_t$$

where q estimates the size of the population index relative to true population size. If CPUE from full catch and effort totals is used as the population index, this becomes:

$$\hat{C}_t = q\hat{N}_t E_t$$

and the parameter q can now be interpreted as a catchability coefficient, defining the proportion of the population harvested per unit effort. Having calculated the full series of fitted abundance and catch values, the best estimates of the initial population size and catchability coefficient can be found by minimising the sum of squared deviations between observed and fitted catches. This **least squares fitting procedure** can be carried out in an Excel spreadsheet using the solver add-in module. Using this function to retrieve known parameters from simulated data is an excellent way to get a feel for how the procedure works, and to understand the limits to its utility. When it comes to analysing real data, the package CEDA (Catch Effort Data Analysis, Section 2.7.1) provides a more sophisticated range of options for fitting models to catch and abundance index data in order to estimate true abundance, including a facility for calculating bootstrap confidence intervals for parameter estimates. Alternatively Bishir and Lancia (1996) provide a maximum likelihood method for catch–effort abundance estimation.

Lancia *et al.* (1996) used this approach to estimate the abundance of introduced wild pigs *Sus scrofa* in Great Smokey Mountain National Park, south-east USA. Park staff control the pig population by intensive trapping and shooting during part of the year, and the catch totals can be used along with effort data to monitor the effectiveness of the cull. The table below shows the weekly catch totals for 1987, along with the hunting effort, measured in man-hours per week, and the resulting catch per unit effort. The table also shows the fitted population sizes and catches, and the squared deviations, at the point of minimum sum of squares. Figure 2.8 give a visual feel for the goodness of fit by comparing observed with fitted catch and CPUE values over time.

Initial population size is estimated at 303 pigs (bootstrap confidence interval 272–366), with an estimated catchability coefficient of $q = 0.00126$. Given a total catch of 266, this suggests 88% hunting mortality over the entire period. In fact, this probably underestimates population size, hence overestimates hunting mortality, as a result of several likely assumption violations. First, the relatively long sampling period means that significant natural mortality may have taken place, violating the closure assumption. Second, hunting occurred during spring, when increasingly dense vegetation may have caused a decline in catchability, perhaps leading to hyperdepletion (Figure 2.7). This may explain the sudden slump in CPUE seen around week seven. Finally, it is very likely that there was individual heterogeneity in catchability. Nevertheless, given the difficulty of estimating abundance by any other means in this case, this approach could be useful as a last resort, so long as the likely bias is understood and consequent uncertainty acknowledged.

Week t	Catch C	Effort E	CPUE C/E	Fitted population $\hat{N}$	Fitted catch $\hat{C}$	Squared deviation $(C-\hat{C})^2$
1	39	108	0.36	303	41.12	4.49
2	46	117	0.39	264	38.81	51.73
3	29	127	0.23	218	34.78	33.38
4	23	89	0.26	189	21.13	3.51
5	24	104	0.23	166	21.68	5.39
6	20	68	0.29	142	12.12	62.06
7	3	95	0.03	122	14.55	133.31
8	9	87	0.10	119	12.99	15.94
9	12	91	0.13	110	12.56	0.31
10	10	93	0.11	98	11.43	2.05
11	14	109	0.13	88	12.03	3.88
12	11	115	0.10	74	10.67	0.11
13	5	66	0.08	63	5.21	0.04
14	11	59	0.19	58	4.29	45.09
15	9	74	0.12	47	4.35	21.61
16	1	72	0.01	38	3.42	5.85
Total catch:	266				Sum of squares:	388.77

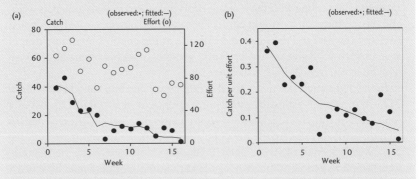

Fig. 2.8 Change in (a) catch and effort, and (b) catch per unit effort over time, during the cull of wild pigs in Great Smokey Mountain National Park. *Source:* Lancia *et al.* (1996).

2.3.5.2 Change in ratios

In this method, information on the proportion of the population observed comes from sampled changes in the **population structure** following a change in that structure. An example of this might be a harvest that targets males so that the population sex ratio is significantly altered, alongside independent visual surveys of sex ratio made before and after harvest (Box 2.6). In order to yield reliable, precise estimates, a **substantial change in ratios** is required between surveys, and this requires that offtake is both substantial and strongly targeted at one of the categories, a constraint that means this method is rarely used.

The key **assumptions** of the basic change in ratios method are:

- The population is closed (apart from offtake);
- The catch is measured accurately;
- Survey detection probabilities are equal across categories.

The first two assumptions also apply to catch–effort methods, and the discussion of these issues in Section 2.3.5.1 is also relevant here. The assumption of **equal survey detection probabilities** across categories is required to ensure that estimated ratios are an accurate reflection of the actual ratios, although methods exist to allow this assumption to be relaxed (Udevitz and Pollock 1991, 1995). The method can also be extended to situations with more than two categories (Otis 1980; Udevitz and Pollock 1991), for example, if several size classes are recognised, or if juveniles

---

Box 2.6 Change in ratios abundance estimation in action.

Given categories x and y (e.g. male and female), numbers removed, C, and numbers seen, n, during surveys 1 and 2, population sizes before and after harvest can be estimated by:

$$\hat{N}_1 = \frac{n_1(C_y n_{x,2} - C_x n_{y,2})}{n_1 n_{x,2} - n_2 n_{x,1}}$$

$$\hat{N}_2 = \hat{N}_1 - C$$

Below is a simulated dataset, based on a population of 1000 before harvest, with 40% males. A reasonably heavy harvest of 20% was applied, with 80% of the offtake males. Ratio surveys before and after harvest observed 25% of the population, encountering each sex randomly. The counts obtained were as follows:

	Male	Female	Total
Harvest, C	160	40	200
Pre-harvest survey, n_1	105	145	250
Post-harvest survey, n_2	61	139	200

The population estimates are therefore:

$$\hat{N_1} = \frac{250 \times (40 \times 61 - 160 \times 139)}{250 \times 61 - 200 \times 105} = \frac{250 \times 19800}{5750} = 861$$

$$\hat{N_2} = 861 - 200 = 661$$

Population confidence intervals estimated by maximum likelihood (using program USER), before and after harvest are respectively 527–3461 and 327–3249. Although the estimated density before harvest is fairly close to the true figure, the precision of the estimates is very low, reflecting the border-line adequate change in sex ratio. The surveys indicate a reduction from 0.42 (105/250) to 0.31 (61/200) males in the population, which is very close to Paulik and Robson's (1969) rule-of-thumb minimum difference of 0.1. This result occurs despite a fairly heavy harvest strongly targeted at males. Effective application of change in ratios methods depends on a change in structure at least as strong as that illustrated here.

cannot reliably be sexed, resulting in three categories (adult male, adult female and juvenile). In principle, the method could also lend itself to the analysis of multi-species offtake, although this has not been tested.

2.3.5.3 Catch-at-age

This heading covers a suite of methods in which data on the age structure of the catch over time are used to **reconstruct the past population size**. To take a very simple example, supposing that 10 individuals in their second year were caught one year, and that this age class was the oldest ever observed, it might be assumed that this catch represents the entire year two cohort. If we know independently that the natural survival rate is 20%, these 10 must have been the survivors from 50 first years in the previous year, and if 100 individuals of this cohort were caught that year, there must have been 150 in existence prior to the offtake. If 20 second years were also caught in that year, the retrospective population estimate would then be 170. This somewhat laborious process, known as **virtual population analysis**, or **cohort analysis**, can be useful in giving some idea of population trends and harvest mortality rates in the past (e.g. Solberg *et al.* 1999; Fryxell *et al.* 2001). However, it relies on the assumptions that the natural mortality rate is known and constant, that there are no individuals alive beyond the maximum age observed, and that there is no migration. Furthermore, in order to reconstruct cohorts that have not yet fully passed through the population (in order to obtain the most recent population size, allowing an assessment of current sustainability), it is also necessary to assume that harvest mortality has remained constant over time. This will clearly be difficult to satisfy in most cases, and is often a huge source of bias in this method.

More complex **statistical catch-at-age** methods have been developed that ameliorate some of these problems (Hilborn and Walters 1992), and these are a mainstay of commercial fishery stock assessments. However, the approach requires a lot of accurately sampled data on the ages or sizes of harvested individuals over a long period of time, which is likely to be prohibitive where the harvest is smaller scale and less intensively monitored. We do not therefore provide details here, but a software implementation and further information are available in MULTIFAN-CL and its associated publications (Section 2.7.1).

2.3.6 Other methods

2.3.6.1 Indirect sign

For animals that are hard to observe, it is often easier to survey their signs, such as dung, prints, burrows, temporary nests or calls. Physical signs are usually relatively easy to survey using plot or distance sampling. However, in order to estimate absolute abundance from these signs, we also need to know their **rates of production and decay** (Box 2.7). This can be achieved by monitoring a sample of live animals to estimate the number of signs produced per unit time, and monitoring a sample of the signs in order to estimate the proportion that disappear per unit time. However, once all of the sampling and observation error is taken into account for sign density as well as production and decay rates, the net result is often **very low precision** for the abundance estimates (Plumptre 2000).

A further problem with this method is that production and decay rates are **difficult to estimate** appropriately. By its nature, the method is applied to animals that are hard to see, making it difficult to quantify sign production rates. For example, while tame elephants can be used to obtain useful defecation rates if they are foraging naturally,

---

Box 2.7 Calculation of density from indirect signs.

Given sign density, $\hat{Y}$, and assuming that rates of production (p) and decay (d) are constant over the study period, animal density is estimated by:

$$\hat{D} = \frac{\hat{Y}d}{p}$$

Each of the three variables that determine the abundance estimate have associated errors that contribute to the uncertainty of the final estimate. The standard error of any estimate derived from a multiplicative combination of several others can be approximated by a process known as the delta method (Seber 1982):

$$SE(\hat{D}) = \hat{D} \sqrt{CV(Y)^2 + CV(d)^2 + CV(p)^2}$$

African forest elephants *Loxodonta cyclotis* have never be tamed, and are too danger-ous to follow in the wild (Hedges and Lawson 2006). Decay rates are easier to esti-mate, but are generally extremely variable in time and space, violating the assumption of constant decay rate. While this problem can be avoided by relaxing the assumption of constant decay rates (Plumptre and Harris 1995), this requires the variation in the decay rate to be tracked over time and space, and the intense effort required for this is unlikely to be feasible in most cases. As a result, there is often a strong temptation to borrow production and decay rates estimated at times and places other than those at which the sign density estimates are made. This introduces huge scope for bias, which, in combination with low expected precision, makes this a technique to use with more than usual care. Returning to the elephant example, the CITES MIKE (Monitoring the Illegal Killing of Elephants) programme suggests that typical defecation rates can be used to estimate Asian elephant abundance in weakly seasonal moist forests because rates have been shown to be fairly constant and consistent across sites, but this approach is not appropriate in strongly seasonal environments, where we only know that the defecation rate is likely to be highly variable (Hedges and Lawson 2006).

In principle, **call rates** might be used as a crude index of abundance for vocal ani-mals, but they will rarely be linearly related to abundance because of a variety of confounding factors, particularly time of day, time of year and weather conditions. While it may be possible to control for these factors by using regression models to examine their relationship with call rate (residual variation then hopefully being directly related to abundance), social facilitation of calls cannot be dealt with in this way. For example, many territorial species, such as the green peafowl discussed in Box 2.19, tend to call more when local densities are higher. When the calls of individuals can be distinguished (e.g. some birds, Gilbert *et al.* 1994), the territory mapping approach (Section 2.3.1) might be used.

2.3.6.2 Presence–absence survey

The primary aim of a presence–absence survey is to estimate **proportional occu-pancy** across a number of sites, as a proxy for abundance. For example, if presence were detected in 10% of sites visited, it would suggest a much lower population size than if 90% of sites were apparently occupied. The **definition of a site** here is case-specific. For example, it may be a discrete habitat island, or, in continuous habitat, a systematically defined grid square. Surveys of this kind can use **any com-bination of cues** to confirm presence, including direct sightings, indirect signs, calls or automatic monitoring systems such as camera traps. This flexibility means that for some species, particularly those that are rare or hard to see, occupancy surveys may provide an efficient means of assessing changes in abundance (Thompson 2004; Joseph *et al.* 2006).

This approach has two main drawbacks. First, it is a relatively **crude index of abundance**. For example, population size at a site is likely to decline substantially well before the population is recorded as absent. Second, the failure to detect a species does not necessarily mean that it is absent; it may simply be hard to detect.

Furthermore, the **probability of detection** given species presence may vary in space and time, giving rise to spurious apparent differences in occupancy.

The problem of variable detection probability can be solved, with some extra effort, by **visiting each site several times**. Just as capture histories provide information on individual detection probabilities in mark-recapture studies (Section 2.3.4), so detection histories can allow site-specific detection probabilities to be estimated. In this way, an estimate of occupancy can be obtained that is not biased by failure to detect presence in some cases (MacKenzie *et al.* 2002; MacKenzie *et al.* 2005). To use this approach, there must be **variation in perceived occupancy**—if sites are found to be occupied either almost always or very rarely, there will be no power to estimate detection probability. The amount of effort required depends primarily on the desired precision and the expected detection probability, however, as a rule of thumb, around 100 sites visited on five occasions is a realistic goal if detection probability is around 0.25 or above. The method assumes that occupancy does not change during the survey, and **repeat visits** should therefore be made over as short a space of time as possible. Defining sites that are large relative to patterns of movement in the target species also helps to ensure that absence on a given visit does not simply mean that individuals have temporarily moved away (in which case, occupancy would reflect space use rather than abundance). Detailed recommendations for designing presence–absence surveys are given by MacKenzie and Royle (2005).

As well as assuming constant occupancy during the survey, the basic repeated-visit occupancy analysis also assumes that the probability of detection given species presence is constant across sites. In order to meet this requirement, it is important to standardise the survey, for example, by searching with **constant effort** at each site on each occasion. Where detection probability still varies, even with carefully controlled survey effort, and this variation is caused by variation in **local abundance**, it is possible to estimate **absolute abundance** using presence–absence data (Royle and Nichols 2003; Royle 2004; Stanley and Richard 2005). This approach works best when numbers at each site are modest, with at least some sites unoccupied, and when the variation in numbers between sites is not great. However, while this method can provide an estimate of the total population size, there is currently no way to define the effective area occupied by that population in continuous habitat. At the time of writing, the method is therefore useful only where sampling units have clear boundaries, within which all individuals are susceptible to detection. Tools for fitting occupancy and abundance models to presence–absence data are available in the software PRESENCE (Section 2.7.1).

2.3.7 Which method is best?

Deciding how best to estimate abundance rests on knowing which analytical methods are available, and balancing the strengths and weaknesses of each (summarised in Table 2.3) with practical constraints such as species characteristics, the working environment, existing data and the financial and labour resources

Table 2.3 Key advantages and disadvantages of the main census methods discussed in Section 2.3.

Analysis type	Field application	Key advantages	Key disadvantages
Complete census		Eliminates statistical uncertainty	Very rarely possible
Plot and plotless sampling	All	Minimises statistical uncertainty (no model error)	Only effective for abundant, static species / objects
	Block plots	Easily applied at small scales	Inefficient at large scales
	Strip plots	Less effort wasted travelling between plots—provides wider coverage	Not effective when travel impairs observation
	Plotless samples	Can reduce required survey time	Assumes random distribution of individuals (very rarely justified)
Distance sampling	All	Not biased by heterogeneous detection probabilities	Requires substantial sighting rate to yield good precision of estimates
	Points	Useful when travel impairs observation	Yields low coverage at large spatial scales
	Lines	Less effort wasted travelling between points—provides wider coverage	Difficult where terrain hinders travel
	Trapping webs	Brings the benefits of distance sampling to species that cannot easily be seen	Very intense trapping effort required; biased by high rates of movement
Mark-recapture	All	Can be applied to species that are difficult to observe; can provide supplementary information on survival and productivity rates	Often biased by heterogeneous capture probabilities; requires a substantial proportion of the population to be observed; sampled area often hard to define

Table 2.3 (*Con't.*)

Analysis type	Field application	Key advantages	Key disadvantages
Mark-recapture (con't)	Live recapture	Requires relatively low technical capacity	Usually labour-intensive; can disrupt behaviour and distribution of the population
	Camera trapping	Eliminates the need for live trapping	Requires natural individual-specific markings (stripes, spots etc.); high start-up costs
	Genetic (hair, dung)	Eliminates the need for live trapping	Requires high technical capacity
Offtake-based	All	Data readily available when harvest well-monitored, minimising effort required	Often biased by heterogeneous capture probability or poor quality data; require heavy depletion
	Catch–effort	Can be developed into full population model if applied over longer periods	Easily biased by inappropriate measurement of effort
	Change in ratios	Data requirements relatively modest	Population needs to be structured into readily observed categories, with harvest strongly skewed
	Catch-at-age	Gives detailed age-structured results	Extremely data-demanding, not widely applicable outside large-scale fisheries
Others	Indirect sign	Can be applied to otherwise invisible species	Great scope for bias in calibrating sign with abundance
	Presence–absence	Useful for species both rarely seen and hard to catch; flexible data requirements	Sampled area usually difficult to define

available. This can be a difficult judgement to make, and every situation is unique, but it is possible to make some **general recommendations**. Figure 2.9 summarises the most important factors that determine the kinds of techniques that are most likely to be applicable in a given situation.

In general, the more complex the model used to estimate the proportion of the population seen, the greater the model error, and the lower the precision will be. On the other hand, making strong assumptions that are difficult to evaluate or satisfy is likely to cause serious bias. We are therefore looking for the simplest possible analytical approach that is practically possible without violating key assumptions. So although complete census is conceptually the simplest possible approach, the assumption that the entire population can be observed is so difficult to satisfy in practice that the method is almost never appropriate. Plot sampling relaxes this assumption by restricting search to more manageable areas, and remains relatively simple, so is the best choice when the species is easily seen, relatively abundant, and static. By adding a little more analytical complexity, distance sampling (line or point transects) can be applied to less visible or more mobile species, and has a relatively low risk of bias. It is therefore a good choice in a wide range of cases, and

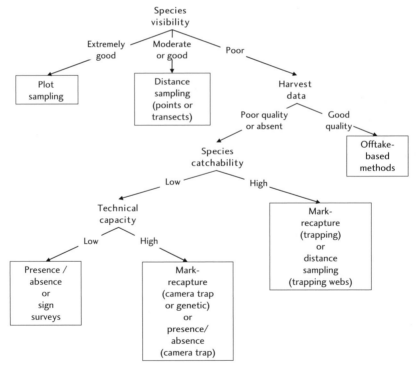

Fig. 2.9 A decision tree for identifying the most appropriate census method for a given situation. This provides a general guide for the most likely approach, but does not rule out innovative uses of apparently inappropriate methods.

should be applied if there is any indication that plots cannot easily be counted with complete certainty. For species that are extremely difficult to observe directly, mark-recapture methods are the most common choice. However, it is often difficult to meet the assumptions of these methods, particularly regarding lack of heterogeneity in capture probabilities, and great care is needed to ensure that field methods are designed to minimise bias. Offtake-based methods might appear to be ideal for exploited species, and may be the only practical option in some cases, although they are the most prone to bias of any of the methods discussed, and should be used with more than usual caution. It will usually be far preferable to estimate abundance independently of harvest if possible.

2.3.8 The future

The methods discussed in this section represent the core of the analytical toolbox, but this is an active area of research, and new methods are constantly being developed. **New technologies** can be applied innovatively to extend the utility of existing census methods. For example, individual animals can be identified from genetic analysis of faecal, feather or hair samples (Rudnick *et al.* 2005; Petit and Valiere 2006), or from camera trap images of species with naturally unique markings (Karanth and Nichols 1998), so allowing mark-recapture methods to be applied. Also, **combinations of analytical methods** are increasingly being developed in order to alleviate some of the existing constraints. For example, Borchers *et al.* (1998) provide a means of applying mark-recapture theory to data from multiple observers to estimate the detectability of individuals on line transects when detection on the line is not certain, while Efford (2004) provides a distance-based method to estimate the effectively sampled area in mark-recapture studies. **Trapping webs** offer the possibility of applying distance-based analytical techniques to trapping data (Jett and Nichols 1987; Buckland *et al.* 2004; Lukacs *et al.* 2005), while **traps or lures** can be used in conjunction with behavioural studies to increase detectability for distance sampling of difficult-to-observe species (Buckland *et al.* 2006). Developments of this kind will no doubt proliferate in the future.

2.4 Measuring demographic rates

Demographic (or vital) rates are the processes that lead to change in population size; births, deaths, immigration and emigration. While population size is a key indicator of the state of the system, a full understanding of biological sustainability also requires an estimate of the population's productive potential. For example, many of the sustainability assessment and prediction techniques introduced in Chapters 4 and 5 require us to know the intrinsic rate of increase for the logistic growth model (Section 1.3.1.1), or rates of productivity and survival, which together determine population growth rate. In this section, we introduce methods for estimating population growth rate, followed by methods for estimating survival (including harvest mortality) and productivity rates. We then briefly discuss methods for defining density-dependent responses in these demographic rates, and touch on the estimation of physical growth and transition between size classes.

2.4.1 Population growth rate

Given two population totals N, counted t years apart, the annual **finite rate of population change** can be estimated by:

$$\lambda = \left(\frac{N_t}{N_0}\right)^{\frac{1}{t}}$$

The parameter λ can be understood as a multiplication rate. For example, if a population of 100 grows by 10% in a year, $\lambda = 1.1$, and the population in the next year will be $100 \times 1.1 = 110$. While this approach quantifies the net change between two points in time, it is more often necessary to estimate the average rate of change over longer periods. In this case, the appropriate method is to regress the natural logarithm of a series of regularly estimated population sizes against time. The slope of this regression gives us the **instantaneous rate of change**, r, which is related to the finite rate of change by:

$$r = \ln(\lambda)$$

The abundance values used in this approach will usually be estimated using one of the closed population methods described in Section 2.3. However, it is also possible to estimate abundance, and hence growth rate, over time using the open population mark-recapture methods described below in Section 2.4.2.2. An appealing feature of these methods is that the estimation of growth rate is an integral part of the model fitting procedure, and software implementations (Section 2.7) therefore provide estimates of growth rate and its associated precision.

A key parameter in simple population models is the **intrinsic rate of increase**, r_{max}, which is the maximum instantaneous rate of population growth at very low density (see Sections 1.3.11, 4.2 and 4.3.3 for more details on the meaning and use of this parameter). There are several ways in which r_{max} can be estimated:

- From a series of abundance measures of a small but growing population;
- From maximal survival and productivity rates;
- From a time-series of catch and effort data;
- From comparisons with other species.

The approach based on a series of abundance measures makes the strong assumption that the population is well below carrying capacity and showing its maximal possible growth. If this assumption can be justified, the **regression method** described above might be used to estimate the growth rate, hence r_{max}.

The second option measures **survival and productivity rates**, again making the assumption that the sampled population is well below carrying capacity, so that the individuals observed are free from competition. Growth rate can then be derived from these maximal rates (Box 2.8). Individuals from a very small population may provide a reasonable approximation for this, although obtaining adequate data from such depleted populations is likely to be a challenge. Captive individuals might also be used, although this approach should be used with great caution. Demography in captivity is unlikely to be equivalent to wild conditions, particularly with regard to survivorship in the absence of natural mortality risks such as predation.

The third option for estimating r_{max}, along with population carrying capacity k, is to fit a density-dependent **population model** to a time-series of catch and abundance data, extending the catch–effort method of abundance estimation introduced in Section 2.3.5.1 (see Section 4.3.3 for details of how to do this). Estimates of r_{max} made using this approach are most robust if at least some estimates of true abundance are available along with catch data, but indirect indices of population size can also be used. The method is most commonly applied to catch and effort data, using catch per unit effort as the abundance index. The minimal data requirements are thus a time-series of total catch along with sub-samples of catch and the effort required to produce it. It is preferable to monitor total effort as well as catch, though, if it can be done without bias, since this minimises sampling error.

Box 2.8 Estimating intrinsic rate of increase from maximal survival and productivity rates.

From age-specific estimates of maximal survival rate, S_j, calculate survivorship (the probability that a newborn will survive to age i):

$$l_i = \prod_{j=1}^{i} S_j$$

Given the maximal age-specific productivity rate, P_i (the number of young females produced per year by a female aged i), and ages at first and last reproduction (respectively a and b), r_{max} can be estimated from:

$$\sum_{i=a}^{b} l_i P_i e^{-r_{max} i} = 1$$

This is a generalisation of Cole's (1954) equation, which has been widely used to estimate r_{max}. However, unlike Cole's equation, it includes survival through the term l_i, which makes it more appropriate for our purposes. There is no explicit solution for r_{max} in this equation, but it can be solved in a spreadsheet by finding the value of r_{max} that brings the sum to 1 (the solver add-in in Excel provides a useful tool for achieving this). Slade *et al.* (1998) provides a range of ways in which simplifying assumptions can make this task easier. The most basic case is when reproduction starts in the first year and continues indefinitely, and survival and productivity rates are constant with age. In this case, r_{max} is given by:

$$r_{max} = \ln(S(1 + P))$$

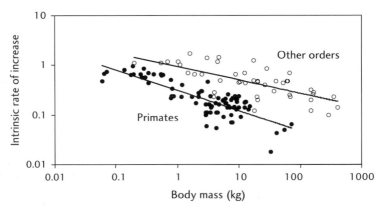

Fig. 2.10 An allometric relationship between body mass, M, and intrinsic rate of increase for mammals. Note the log scales. For primates, $r_{max} = 0.31M^{-0.41}$; for other mammals, $r_{max} = 0.94M^{-0.27}$ (Rowcliffe *et al.* 2003).

Finally, in the absence of specific information, a possible alternative is to estimate r_{max} by **comparison** with a similar but better known species. In vertebrates, intrinsic growth rate is related to size, with larger species generally having lower rates. This allows the comparative approach to be generalised, using allometric equations to predict r_{max} (Figure 2.10). However, this approach is limited by the low precision of its predictions—among the 127 species represented in Figure 2.10, fitted values differed from actual estimates up to three-fold.

2.4.2 Survival rate

Survival is one component of population growth, and as we saw in the previous section, it can be used to estimate growth rate. It can also be used in predictive models (Chapter 5), and is itself affected by harvest mortality, which can be an important indicator of sustainability. Survival rate expresses the probability that a given individual will survive over a given period of time, and for most practical purposes, it can be expressed as a finite rate. Thus, if the annual survival rate in a population is 0.6, on average 60% are expected to survive each year. Survival can be estimated either from population count data (Section 2.4.2.1), or from a sample of known individuals (Section 2.4.2.2).

2.4.2.1 Using count data

Population ratios

In principle, survival rates can be estimated as ratios from repeated estimates of population size split by age (Box 2.9). For example, if it is known that a population numbered 100 individuals of all ages (aged 0 and older) in one year, and the next year there were 80 adults, the survival rate would be estimated as 80/100 = 0.8.

Box 2.9 Estimating survival rate from population estimates.

In the simplest case, in which numbers of young of the year, N_0, and older, N_{1+}, individuals are estimated at two or more points in time (usually annually), time-specific but age-independent survival rate, S_t, is given by:

$$S_t = \frac{N_{1+,t+1}}{N_t}$$

However, in long-lived species, demographic rates may vary greatly with age. Age-specific variation in survival rate can be quantified by generalising the above equation for any age class, a:

$$S_{a,t} = \frac{N_{a+1,t+1}}{N_{a,t}}$$

Because older age classes are generally scarce, and survival rate often reaches a roughly stable value after a period of maturation, it will usually be both sensible and reasonable to define a maximum age class, A, beyond which survival is assumed to be constant. In this case, the survival rate for individuals aged between A and the maximum recorded age, A_{max}, is:

$$S_{A+,t} = \frac{\sum_{a=A+1}^{A_{max}} N_{a,t+1}}{\sum_{a=A}^{A_{max}-1} N_{a,t}}$$

The problem with this approach is that it relies on very accurate and precise age-specific population estimates, which will rarely be achievable. A more practical approach in some cases might be to quantify only total population size, and carry out independent surveys of age-structure based on a sampling method that takes account of any differences in detectability between age classes. However, this adds further sampling error to the calculation. Relatively small amounts of error in the constituent numbers propagate into considerable error in survival rate, which has a strong influence on the assessment of sustainability. Given the low degree of precision yielded by most methods for estimating abundance and age structure, estimates of survival rate based on this approach should be treated with caution.

Static life tables

Given the difficulty of obtaining precise age-structured population estimates, count-based calculation of age-specific survival rate more commonly uses a sample of the population to define age structure. Samples may be taken visually if age classes are sufficiently obvious in the field, but otherwise depend on the retrieval of dead individuals, usually through harvest. Methods for determining age depend on the species. For example, patterns of tooth wear and eruption (Hewison *et al.* 1999; Gipson *et al.* 2000) or annual rings in teeth (Bodkin *et al.* 1997; Costello *et al.* 2004) may be used in many mammals, plumage characteristics are used in many birds (Prince *et al.* 1997), annual rings in scales or ear bones (otoliths) are used in some fish (Buckmeier and Howells 2003; Rifflart *et al.* 2006), and annual rings in woody plants in seasonal environments. A picture of age structure may be taken as a single snapshot in time, or accumulated over time if this is necessary to provide a sufficiently large sample.

The analytical framework for data of this kind is the static life table. The underlying concept here is the same as that behind the calculations for full population estimates outlined in Box 2.9; in essence, you just need to substitute sample counts, n_a, for complete population estimates N_a (Box 2.10). Skalski *et al.* (2005b) provide a comprehensive set of statistical methods for static life tables. The key assumptions of this approach are that:

- Population size is stable;
- The age structure is stable;
- Ages are accurately estimated;
- Individuals of all age classes are equally likely to be sampled.

The **stable population** assumption can be relaxed if the rate of change has been constant in the recent past and is known. Given finite rate of change λ (Section 2.4.1), the survival rate can be calculated as:

$$S_a = \lambda \frac{n_{a+1}}{n_a}$$

While this allows for trends in population size, it still assumes that the **age structure** is stable. Unfortunately this will often not be the case in exploited species. If harvest has just begun, has recently been substantially reduced, or is highly variable, age structure may show strong fluctuations. This will also occur in populations that fluctuate naturally, for example because of highly variable recruitment patterns. In any of these cases, variable age structure leads to biased age-specific survival rate estimates. When faced with this problem, age samples pooled over an extended period, rather than taken from a single snapshot, can help to smooth out fluctuations and reveal the underlying survival patterns. A more robust approach has been developed by Udevitz and Ballachey (1998), using data from both the living population and the harvest sampled over time, and allowing age structure to fluctuate, so long as the population growth rate is independently known.

The third and fourth assumptions (**accurate aging** and **equal detectability**) are necessary to ensure that the age sample faithfully reflects the actual age structure in the population. In practice, this is frequently very difficult to achieve. Apart from the obvious difficulties in obtaining accurate ages in many species, different age classes are likely to have different detectabilities. In the case of visual samples, this might be controlled for by using one of the census methods in Section 2.3 to quantify and control for the variation in detectability. However, when sampling

Box 2.10 Static life table analysis of survival from white-tailed deer harvest.

The table below shows the numbers of white-tailed deer in each age category killed by hunters in one season in part of Michigan, USA (Eberhardt 1969). Reasonably precise survival estimates are obtained for the first three age classes, but the confidence intervals rapidly become very wide as age increases. Given that there is no evidence of systematic changes in mortality rate after this time, it may be sensible to calculate a single survival rate for deer aged over 5.5 and older, i.e. $19/35 = 0.54$. Note that the youngest possible age category (0.5) does not appear because this group is known to be under-represented in hunting offtake. In order to estimate the full survivorship pattern, alternative methods would be needed to estimate rates of mortality between 0 and 1.5 years of age. The standard error is estimated using the approach described by Skalski *et al.* (2005b, p. 162).

Age a	Number sampled n_a	Survival rate $S_a = n_{a+1}/n_a$	Standard error $SE(S_a)$
1.5	425	0.64	0.05
2.5	274	0.54	0.06
3.5	149	0.36	0.06
4.5	53	0.32	0.09
5.5	17	0.47	0.2
6.5	8	0.75	0.41
7.5	6	0.5	0.35
8.5	3	0.33	0.38
9.5	1	1	1.41
10.5	1		

harvested carcasses, the data are often irretrievably biased. For example, some age classes may be more vulnerable (e.g. naïve young geese may be more prone to being shot; Wright and Boyd 1983), or harvesters may show active preferences (e.g. fishers adapting their gear to target larger fish; McClanahan and Mangi 2004). A common response to the latter problem is to estimate survival only for the ages where there is assumed to be no selectivity (Box 2.10), but this assumption cannot usually be tested and is therefore risky.

Thus static life tables are effective only when high quality, unbiased data on age structure are available, alongside supplementary information on population trends. It is also important to have a **large sample size**, as small samples lead to problems with counts that do not decline continuously with age (leading to survival rates greater than 1) and extremely imprecise estimates (Box 2.10). As a rule of thumb, a basic life table analysis is likely to require a sample of at least a thousand individuals to provide reasonably precise survival estimates (standard error <0.1 up to the fourth age category).

2.4.2.2 Using marked individuals

There are two basic forms of mark-recapture that can be used to estimate survival rates. First, individuals may be recaptured or resighted alive at discreet intervals. Second, individuals may be recovered dead, often through harvest. It is also possible to use both types of information in combination. Williams *et al.* (2002) and Amstrup *et al.* (2006) provide comprehensive details on the current state of the art in these methods. The methods are implemented in widely available free software (especially MARK; Section 2.7.1), and we refer the reader to these programs and their associated literature for detailed advice on the their practical application (in particular, the online introduction to MARK by Cooch and White 2006). Here we outline the basic concepts, and briefly discuss the most important considerations relevant to designing a field study.

Live recaptures

Survival analysis based on recapture or resighting is often referred to as the Cormack–Jolly–Seber (CJS) model, after those who pioneered the approach. It works by using information over **several discrete observation occasions** to separate the probabilities of observation and survival. Imagine that 100 animals are caught and marked, and that during a second capture session, some time later, 30 of them are seen again. This indicates that the probability of being both alive and observed is 0.3, which could mean anything between 100% survival rate with 30% chance of being seen, and 30% survival rate with all of the survivors being seen. At this stage we can say no more than this. Suppose that a third capture occasion produces 18 individuals that were marked on the first occasion, only nine of which were also seen on the second occasion. This suggests that the probability of observing a marked individual was 9/18 = 0.5. Given that the probability of being both alive and observed is the product of the survival and observation probabilities, the survival rate can now be estimated as 0.3/0.5 = 0.6. The basic data requirement for this type of survival analysis is therefore a set of individual capture or observation histories over three or more occasions.

The method is closely related to the closed population mark-recapture method of abundance estimation, and apart from population closure, the **assumptions** of that method also apply here (see Section 2.3.4). In addition, mark-recapture survival analysis assumes that:

- The marked individuals are a random sample of the wider population;
- Sample occasions occur more or less instantaneously relative to the gaps between them.

Random sampling of marked individuals is necessary if the results are to be assumed **representative** of the population as a whole. If the capture method is biased towards certain subsections of the population, inference about the wider population will be impossible. The most worrying example of this is when the capture or marking procedures themselves create a sub-population by reducing survival probabilities, either directly (e.g. McCarthy and Parris 2004), or because the marks make individuals more vulnerable to harvesters. In some cases it may be possible to detect and control for such effects (Box 2.11), but they should ideally be avoided altogether. Capture methods that seek to maximise returns, such as baiting, should be used with caution because they are likely to yield unrepresentative samples if classes of animal respond differently to the bait. A final important

---

Box 2.11 Survival estimation in action: mark-recapture analysis of crayfish.

Box 2.4 illustrated closed population mark-recapture analysis in a study estimating crayfish abundance in Madagascar (Jones and Coulson 2006; Figure 2.11). That analysis used data from a single five-visit survey. However, these surveys were repeated four times over a 22-month period, allowing rates of survival between surveys to be estimated. A range of models were fitted to the data, allowing survival and capture probabilities to vary with size class, sex and time. The number of times each crayfish was caught was also fitted as an individual covariate in order to look for possible effects of capture on survivorship.

The most fully parameterised models fitted the data well, suggesting that any heterogeneity in survival and capture probabilities was explained adequately by the covariates used. The best fitting model showed generally lower survival rates in smaller crayfish, and strong evidence for a negative effect of handling on survival in large crayfish. Controlling for this handling effect, large crayfish survival rate was estimated at over 0.7, but the estimate was only around 0.4 if this effect was ignored. This emphasises the need to take seriously the possibility that research activity can have a strong effect on the parameters it attempts to estimate.

Fig. 2.11 The Malagasy crayfish *Astacoides granulimanus*, found in forest streams, and harvested for food. They can also be affected by temporary capture; repeated handling dramatically reduces their survival rate. Photo © Julia Jones.

consideration required to achieve representative samples is that the geographical location of captures or observations should be well spread throughout the population of interest.

Satisfying the fourth assumption (**instantaneous sampling**) needs careful consideration of the duration of recapture or resighting events. Survival estimates are likely to be biased if the recapture period is extended and coincides with a period of substantial mortality, particularly if survival rates vary strongly over time. The ideal is to restrict recaptures or re-sightings to a very short period of time relative to the intervening periods. Directed recapture or resighting effort should therefore be used in preference to opportunistic observations. In practice, extended capture periods may be unavoidable, since robust survival estimation also requires a reasonable recapture probability (as a rule of thumb, at least 0.2), and this may be achievable only through prolonged effort. In some cases, particularly when survival rate is relatively constant over time, bias caused by extended capture periods may be minimal (O'Brien *et al.* 2005). Alternatively, for seasonal species, the risk of bias can be minimised by timing recapture events to occur over a part of the year in which mortality is thought to be minimal.

A final implicit assumption to note is that the estimated survival rate reflects mortality and not **emigration**. In practice, the rate will often reflect a combination of these processes, and it is often referred to as **apparent survival** to reflect this uncertainty. Methods that can be used to separate true survival from migration are introduced in Section 2.5.1.

Most commonly, mark-recapture survival analyses use **annual spacing** of captures; however, it may be sensible to adjust this. First, it may be useful to estimate survival rates during different parts of year, for example, comparing hunting and non-hunting seasons in order to assess the impacts of hunting. In this case, the gaps between recapture events need not be of equal length. Second, the model fitting procedures work best when survival rates are reasonably high (greater than around 0.4) but not too close to 1, and this may require the spacing of recaptures to be adjusted so that survival rates in each interval meet this requirement. As an extreme example, in very short-lived species, no individuals may survive from one year to the next, in which case several recapture events will be required during the year.

Dead recoveries

Just as the mark-recapture method enables survival rate to be separated from recapture rate, the analysis of marked individuals **recovered dead** separates survival rate from the probability of dying and being reported (the recovery rate). Imagine that a group of 1000 animals is marked, and that 100 and 50 of these are recovered in each of the following two years. The fact that the numbers recovered halved from 100 to 50 indicates a survival rate of 0.5. Patterns of marking and recovery can thus be used to extract information about survival rate. The basic data required for this analysis are the timing of marking and (if it happens) recovery for each individual.

Dead recovery analysis is particularly suited to exploited species because harvest can be an excellent source of recoveries. On the face of it, it may seem that only harvest mortality is accounted for in this case, but this is not so. Regardless of the source of recoveries, the method estimates overall survival rates, including both natural and harvest mortalities (see Section 2.4.2.4 for ways to separate these out). The **assumptions** of the method are essentially the same as for the live recapture approach, and the above discussion of these assumptions also applies here. The main difference is that, although the marking occasions should be as brief as possible, and take place at a time when there is minimal mortality or movement, the recoveries need not be restricted in time. Also, if **age-specific variation** in survival rate is to be estimated, it is essential to mark a combination of both adults and young—marking young alone will not provide the necessary information.

The **sample size** required to obtain an estimate with a given degree of precision also depends on the expected survival and recovery rates. As an approximate rule of thumb, if the average survival rate is to be estimated for a study period with six marking occasions, and the probability of a dead individual being reported is expected to be around 0.1, to achieve a coefficient of variation of around 10%, a few hundred individuals would have to be marked on each occasion if the survival rate were high (around 0.8), rising to a few thousand individuals per occasion if the survival rate were low (around 0.2). Increasing the number of parameters to be estimated (for example, by estimating time- or age-specific survival rates) increases the required sample size by a factor of at least two, whereas improving the reporting rate can give an equally dramatic reduction in the sample size required. Software BAND2 (Section 2.7.1) provides a tool for estimating sample sizes required to obtain a given level of precision and a given set of conditions.

Known fates

When individuals are marked and released, usually only a small proportion of them are found when they die. The dead recovery analyses described above enable us to deal with this situation. However, sometimes it is possible to follow fates perfectly, such that the time of death is known, at least approximately, for all marked individuals. For example, this may be the case for highly visible or sessile species that can easily be observed directly, or for animals that can be located at will using radio-tags (e.g. Nybakk *et al.* 2002). Ideally, all individuals are of known age at the time of marking, as this allows age-specific survival rates to be estimated. One way to achieve this might be to follow individuals from birth.

The simplest way to analyse data of this kind is to treat them as a life table. The method in Box 2.9 can be used to calculate the survival rate for each period, using the numbers of marked individuals known to be alive after each time interval since marking instead of sample counts. This approach is known as a **cohort life table**. Unlike static life tables, no assumptions are made about the population rate of change.

A more flexible approach to analysing known fate data is to use 'time-to-event' **survival analyses** such as the Cox proportional hazards, Kaplan–Meier or parametric survival models (Smith 2005; StatSoft n.d.). These approaches estimate mortality rates from the precise times between marking and death for a sample of individuals. One benefit of this approach over the life table is that it is easy to include in the analysis individuals that are followed for a period but not observed to die before monitoring ceases, so long as a reasonable number of deaths are actually observed. More importantly, these methods are forms of regression analysis that allow the influence of possible explanatory factors on survival rate to be formally tested. In common with mark-recapture methods, survival rate estimates based on known fates are only **representative** of the wider population to the extent that the individuals observed are representative. The usual care in selecting study subjects is therefore required.

2.4.2.3 Which method is best?

The range of methods available for the estimation of survival is as wide as that for abundance estimation; however, the choice of method will usually be much more restricted by the type of data that can be obtained from your organism (Figure 2.12). Where there is a genuine choice, individual-based methods should be preferred, since they generally provide the most robust way to obtain unbiased estimates of survival, and allow testing of hypotheses about the factors that may influence it (such as harvest pressure).

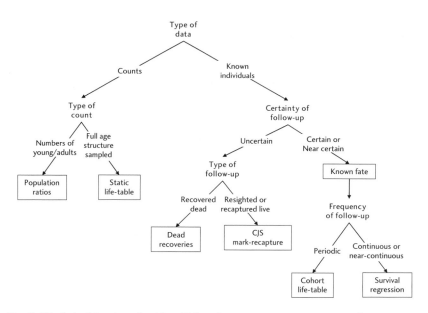

Fig. 2.12 A decision tree for identifying the most appropriate survival estimation method for a given type of data.

2.4.2.4 Harvest mortality

The survival analysis methods described above provide estimates of overall survival rate, which encompasses both natural and harvest mortalities. If we wish to assess the impact of the harvest, it is useful to know what proportion of the overall mortality it represents (see Chapters 4 and 5). Harvest mortality can be measured either by monitoring directly the amount of offtake and relating this to the population size, or by using one of several indirect approaches.

The ratio of catch to pre-harvest population size gives an obvious direct measure of harvest mortality (Box 2.12). For example, Menu *et al.* (2002) used this approach to estimate harvest mortality in greater snow geese *Anser caerulescens* in North America, taking total harvest from national hunter surveys, and estimating pre-harvest (autumn) population size from a combination of spring aerial photo surveys and summer survival rate estimated from ring recovery data. A major challenge with this approach is to obtain an accurate measure of the catch. Harvest may be monitored in many ways, from accompanying harvesters in order to observe offtake directly, through monitoring of collection points such as hunting bases or markets, to various forms of questionnaire to provide indirect reports on offtake (see Chapter 3 for guidelines on these field methods). When applying these methods to measure total offtake, it is important to recognise the potential bias that may arise. For example, the more harvesters are involved, and the more remote the monitoring point is from the point of offtake, the more likely it is that significant amounts of offtake will go unrecorded. In this case, any pretence at complete coverage should be abandoned, in favour of a rigorous **sampling approach**.

Box 2.12 provides analytical methods for calculating total offtake from a sample. The crucial assumption with these methods is that the harvesters and occasions sampled are **representative** of the overall pattern of harvest. When sampling occasions, it is usually easy to ensure representative coverage through a carefully designed sampling regime, for example, by picking random days of the week, or using a regular cycle that covers all days of the week equally. However, it is often more difficult to achieve a representative sample of harvesters. This is because harvesters that are less accessible to monitoring are likely to have different patterns of effort or success than those that are easier to reach, particularly if the survey relies on voluntary responses. The best option is to start by identifying the entire harvester population, then ensure that all individuals have an equal chance of being monitored through stratified or strictly random sampling. Where this is not possible, it may be necessary to make a special effort to sample individuals not covered by the main survey, for example, by following up voluntary returns with direct observation of non-respondents. This can make it possible to detect and correct for any bias in the main survey.

However, it is important to bear in mind that even a well-designed, representative offtake sample may suffer from under-reporting. This is particularly true if there are social or legal pressures against harvest, but can be a problem even in the

Box 2.12 Direct estimation of catch and harvest mortality.

If the total harvest, C, and pre-harvest population size, N, are known or can be estimated, harvest mortality is:

$$\hat{M} = \frac{C}{N}$$

$$\text{SE}(\hat{M}) = \hat{M} \sqrt{\text{CV}(C)^2 + \text{CV}(N)^2}$$

The offtake-based methods for estimating population size (Section 2.3.5) are a natural way to obtain the necessary data for this since they require offtake to be known, and provide estimates of population size. The precision of estimated harvest mortality will obviously be higher if the total harvest can be recorded without error ($\text{SE}(C) = 0$). Unfortunately this will often not be possible, in which case offtake must be sampled instead. If a proportion, p, of the total harvest is sampled, giving a catch of c, estimated total catch is given by:

$$\hat{C} = \frac{c}{p}$$

$$\text{SE}(\hat{C}) = \frac{\text{SE}(c)}{p} \sqrt{1-p}$$

where $\text{SE}(c)$ is derived from the empirical sampling variance of the catch. Sampling catch can be done either by direct observation, or by asking harvesters to report their offtake to you. In the case of **direct observation**, sampling can be done in a number of ways, recording either the total offtake across all occasions for a sample of harvesters, recording the total offtake across all harvesters on a sample of occasions, or, less desirably, recording the offtake from a sample of harvesters on a sample of occasions. Whichever method is used, the sampling proportion is calculated as the ratio of observed to total possible effort: $p = e/E$, where the definitions of effort depend on the approach taken:

Sampling approach	Effort observed, e	Total possible effort, E
Occasions	Number of occasions, t	Total number of occasions, T
Harvesters	Number of harvesters, h	Total number of harvesters, H
Both occasions and harvesters	Number of harvester occasions, $\sum\limits_{i=1}^{t} h_i$	Total harvester occasions, $T\bar{H} = \dfrac{T\sum\limits_{i=1}^{t} H_i}{t}$

For example, if 20 harvesters were monitored on 100 days over a year, during which the average total number of harvesters active was 60, the sampling effort would be $(100 \times 20)/(365 \times 60) = 2000/21900 = 0.09$.

In the case of **self-reported harvest**, the sampling proportion, p, is estimated as the probability that any given harvest will be reported. This probability, the **reporting rate**, can be estimated by following up the initial voluntary survey with a direct check on offtake to find the proportion of harvest that was not reported. For example, this approach is used to monitor game species in North America in situations where many of the animals are individually tagged (Rupp *et al.* 2000). Registered hunters are contacted by post, requesting the numbers harvested as well as any tag identifiers, while direct checks on offtake are also carried out to detect unreported tags. Skalski *et al.* (2005b) and Pollock *et al.* (1995) provide further details on this and other methods for estimating reporting probabilities.

absence of such pressures if respondents have little motivation to participate in the survey, or if they simply have poor recall. Less often, offtake may be exaggerated if harvesters see large catches as prestigious, or if harvesters that work in groups each independently report their group's catch.

It is not always necessary to know the catch to estimate harvest mortality. For example, if harvest takes place over a relatively short space of time, during which there is no significant natural mortality, harvest mortality rate can be calculated from **population abundance estimates** before and after harvest (Box 2.13). A downside with this approach is that abundance estimates usually suffer from low precision, and the combination of two imprecise estimates leads to extremely imprecise estimates of mortality rate. Where it is impossible to get reliable abundance estimates, this approach could in principle be taken with **population indices**, such as basic encounter rates. However, this requires the strong assumption that the index remains linearly related to abundance throughout. It is likely that this assumption will often be violated, leading to biased results.

Where natural mortality during the period of harvest is an issue, we need to disentangle this from the harvest mortality. The dead recovery form of mark-recapture survival analysis described in Section 2.4.2.2 provides a means to do this (Box 2.13). In this approach, if survival rate has been estimated from harvested individuals alone, the recovery rate is the probability that an individual is both harvested and subsequently reported. If we can estimate independently the reporting probability (see Box 2.12 for possible approaches to this problem), we can therefore derive the harvest mortality rate. For example, Calvert and Gauthier (2005) used this approach to estimate seasonal harvest mortality rates in greater snow geese *Anser caerulescens*.

Box 2.13 Indirect estimation of harvest mortality.

When natural mortality is negligible: using population estimates. Given population estimates, N, (or population indices linearly scaled with abundance) before and after harvest, assuming no natural mortality over the interval, harvest mortality is given by:

$$\hat{M} = 1 - \frac{N_{\text{after}}}{N_{\text{before}}}$$

$$SE(\hat{M}) = \hat{M} \sqrt{CV\,(N_{before})^2 + CV\,(N_{after})^2}$$

When natural and harvest mortality operate simultaneously: using dead recovery data.
Given recovery rate, v (the probability of dying and being reported), estimated from harvest returns alone, and reporting rate, p, the harvest mortality rate is given by:

$$\hat{M} = \frac{v}{p}$$

$$SE(\hat{M}) = \hat{M} \sqrt{CV(v)^2 + CV(p)^2}$$

Care may be needed with this approach where harvesters kill significant numbers without recovering them. If this happens and the estimated reporting rate does not take this into account, effective harvest mortality will be underestimated.

2.4.3 Productivity

While survival describes the rate at which individuals leave a population through death, productivity defines the rate at which new individuals enter through reproduction (birth, seed set, etc.). Along with survival, productivity is thus a key determinant of population growth and response to exploitation (Jennings *et al.* 1998). Productivity can be expressed in a number of related ways, depending on the data available and the question one is attempting to answer. Here we define it as the **number of young produced per adult per unit time**. Another common formulation for dioecious species is **fecundity**, the number of young females produced per female. Fecundity is frequently used in population models when it can be assumed that access to males does not constrain reproduction (see Milner-Gulland *et al.* 2003, for an example of the limits to this assumption). The general methods for productivity described below apply equally to the estimation of fecundity, simply requiring data to be restricted to females.

Another important concept in the measurement of productivity is **recruitment**, defined as the age at which individuals functionally enter the population. In population dynamic studies, this is frequently defined as the age at first breeding, while in exploited species it may be defined as the age at which individuals first become vulnerable to capture. For the discussion of methods here, we define recruitment more generally as the point at which reproductive output is measured and used, which will vary from one study to the next. In some cases, offspring may be counted when they are a year or more old, while in others observations may take place at a very early developmental stage (e.g. eggs, embryos or seeds). In this case, in order to obtain a productivity estimate that can be used in the modelling approaches described in Chapter 5, it may be necessary to multiply early stage productivity by the rate of survival to the point of recruitment as defined in the model. However, early survival rates are usually difficult to measure, and add another level of sampling error, so it is best to estimate productivity directly if possible.

2.4.3.1 Using count data

A direct measure of productivity is given by the ratio of young to adults in the population (Box 2.14). The necessary data can be obtained in one of three ways, depending on the practicalities of different field methods:

- Using population abundance estimates separated by age;
- Using sample counts of the population separated by age;
- Using counts of offspring, eggs, embryos, placental scars (Martin *et al.* 1976), fruit or seeds associated with a sample of potentially breeding individuals.

Box 2.14 Measuring productivity using count data.

Given estimated or counted numbers of offspring, y, and adults, a, productivity is simply the ratio of the two:

$$\hat{P} = \frac{y}{a}$$

The precision of the estimate depends on the sampling method. Where y and a represent abundance estimates with their own associated variances:

$$SE(\hat{P}) = \hat{P} \sqrt{CV(y)^2 + CV(a)^2}$$

Where a sample of n individuals is counted without direct connection between individual breeders and their offspring:

$$SE(\hat{P}) = \sqrt{\frac{\hat{P}(1 + \hat{P})^2}{n}}$$

Finally, where n individuals each produce y offspring, the population variance is used (see Box 2.1):

$$SE(\hat{P}) = \sqrt{\frac{s^2(y)}{n}}$$

In the last two cases, if it is known that no individuals have been counted more than once (sampling without replacement), and the total population, N, is also known, the standard error can be reduced according to the proportion of the population not sampled:

$$SE(\hat{P}) = SE(\hat{P}) \sqrt{1 - \frac{n}{N}}$$

Note that these methods can be used to estimate age or sex specific productivity rates in cases where such structure is likely to be important (see for example Milner *et al.* 2007), simply by applying the equations to data separated by age or sex. They can also be adapted to the estimation of other population ratios, such as sex ratio.

When either of the sampling approaches is used, it is, as always, important to ensure your sample is **representative** of the population as a whole. A common cause of bias arises if non-breeding individuals are segregated from the breeding population, making it difficult to ensure that they have an equal chance of being sampled. In this case, the best way to obtain an unbiased estimate may be to calculate first a productivity rate solely for the breeding portion of the population, then multiply this by an estimate of the proportion of the population breeding.

2.4.3.2 Using marked individuals

Because population size from one year to the next is a function of survival and productivity rates, and there is a strict theoretical relationship between these variables, if you know survival rate and population size over a period of time, you can derive productivity rate. The mark-recapture methods described earlier in this chapter are ideal for this, and in fact methods have been developed that build the estimation of productivity into the analysis of mark-recapture data. These methods fall broadly into two categories, robust design, and Jolly–Seber and related methods.

Robust design (Pollock 1982; Williams *et al.* 2002) combines open (Section 2.4.2.2) and closed population (Section 2.3.4) mark-recapture methods, using the former to estimate population size, the latter to estimate survival, and deriving recruitment from these. This requires a design in which periods of closely spaced capture occasions are interspersed with longer gaps. The population is assumed to be closed during each set of close-spaced occasions, but can undergo turnover between each set.

Jolly–Seber and related methods (Jolly 1965; Seber 1965; Williams *et al.* 2002) are essentially extensions of the open population methods used to estimate survival rate, described in Section 2.4.2.2. If we can reasonably assume that the marked portion of the population is a representative, random sample of the whole, the estimated recapture probability can be interpreted as an estimate of the proportion of the population observed on a given occasion. If, in addition to this, we record the numbers of unmarked as well as marked individuals seen or captured, we can then estimate population size, and hence recruitment, as well as survival.

The key **assumptions** for these approaches, and the practical considerations for sampling to provide the necessary data, are largely the same as those highlighted in Sections 2.3.4 and 2.4.2.2. In addition, the assumption of representative sampling required by Jolly–Seber methods necessitates a careful field design to ensure that marked individuals are not targeted for recapture or resighting, either deliberately or inadvertently. Also, when recording the numbers of unmarked individuals for this approach, it is important that there is no possibility of double-counting. Jolly–Seber methods generally result in productivity estimates with low precision, and which are vulnerable to bias due to heterogeneous capture probabilities. In addition, Jolly–Seber methods are generally unable to estimate key parameters for the first and last occasions of a survey.

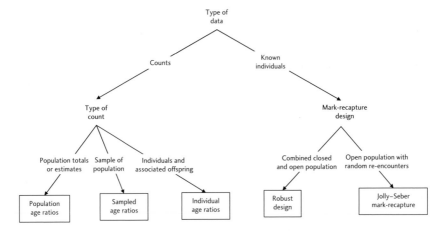

Fig. 2.13 A decision tree for identifying the most appropriate productivity estimation method for a given type of data.

2.4.3.3 Which method is best?

When estimating productivity, the main choice is a relatively simple one between count-based or individual-based methods (Figure 2.13). Appropriate survey timing and careful representative sampling can make count-based methods very effective, and these are the obvious choice for relatively visible species. In species that are more difficult to count or follow up reliably, extensions of mark-recapture methods can be used, and the choice among these depends primarily on the resources that can be devoted to sampling. While more complex and intensive sampling is required for the robust design method, it can greatly reduce the risk of bias and poor precision inherent in Jolly–Seber and related methods. Robust design is therefore preferable if resources allow.

2.4.4 Density dependence

In Chapter 1, we introduced density dependence as a central process for exploited species. In principle, the strength and form of density dependence have a crucial influence on the impacts of harvest, and defining the process can therefore be very helpful in understanding and predicting outcomes. One approach, covered earlier in this chapter, is to assume that the **logistic model** applies. Because this model is defined by a linear decline in growth rate with increasing population, we can fully parameterise it with only two parameters, r_{max} and K, and these can be estimated either independently (Sections 2.4.1 and 2.3 respectively), or jointly by fitting the model to a series of catch and population index data (Section 2.4.1). This approach can be seen as a parsimonious default option, fitting the simplest possible density-dependent model to the available data. The benefit of this is that it allows us to work with cases where information on the precise form of density dependence is lacking, which is a common situation.

The risk in assuming logistic growth is that the factors regulating population growth actually act in significantly **non-linear** ways, potentially leading to very different responses to harvesting. For example, it is thought that some species tend to show little regulation at low population size, but strong regulation around carrying capacity, while others tend to show the strongest regulation when their populations are small (Fowler 1981; Sibly *et al.* 2005), which may be due to differential susceptibility to **competition** among different age or size classes (Owen-Smith 2006). In order to detect and quantify these complications, we require demographic rate estimates (either population growth rate or its constituent parameters, survival and productivity), replicated in time or space at a wide range of different abundances, allowing us to define the demographic response to abundance (Box 2.15). Furthermore, it is highly desirable to separate demographic rates by

---

Box 2.15 Some useful density dependence functions.

In principle, the response of demographic rates to density might take any form. In practice, a few relatively flexible functions have become commonly used, preferred for their wide applicability. Where population abundance, N, and growth rate, $r_t = \ln(N_{t+1}/N_t)$, measures are available, a theta-logistic model can be fitted to the data:

$$r_t = r_{max}\left(1 - \left[\frac{N_t}{K}\right]^\theta\right)$$

This adds an additional parameter, θ, to the logistic model, allowing non-linear growth responses (Fowler 1981). When $\theta < 1$, regulation intensifies as the population declines below K, when $\theta > 1$, regulation is most intense at around K, while $\theta = 1$ gives standard logistic growth. When $\theta = 1$, a simple linear model of r against N can be used, the intercept giving an estimate of r_{max}, and the intercept divided by the slope estimating K. Otherwise, a non-linear fitting procedure is required to estimate the parameters.

Where data are available on either productivity or survival rates (jointly denoted ϕ), functions commonly used to define density responses are the Ricker function:

$$\phi = \phi_{max}e^{-bN}$$

or the Beverton-Holt function:

$$\phi = \frac{\phi_{max}}{1 + bN}$$

In both cases ϕ_{max} defines the maximum survival or productivity rate at low population size, and b defines the rate of decline with increasing density. Where these functions are used to relate recruitment to the abundance of breeders, the Ricker function is more appropriate if the outcome is primarily affected by the density of

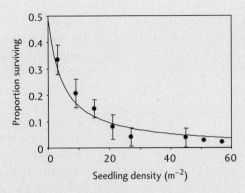

Fig. 2.14 The growing heart of the palm *Euterpe edulis* is cut to provide the Brazilian delicacy palmito, killing the plant. In order to model the impacts of harvest, it is important to define the crucial density dependent processes in a population, achieved in this case by fitting a Beverton-Holt function to seedling density and survival data.

Source: Silva Matos *et al.* (1999). Photo © Henrik Balsler

the breeders (for example, through cannibalism or competition for spawning sites), while the Beverton-Holt function is more appropriate if the density of juveniles affects their own mortality rate (for example, through direct competition for food). The functions can also be applied broadly to any survival or productivity rate and abundance data. For example, Silva Matos *et al.* (1999) studied the demography of a harvested palm, *Euterpe edulis*, in the Atlantic forest of Brazil, finding that the key point of density-dependent regulation was seedling competition (Figure 2.14). They found this by monitoring the density and survivorship of seedlings in 100 1 m^2 plots, with densities ranging from 0 to >50 m^{-2}. A Beverton-Holt function was fitted to the seedling survival data from the plots, giving a maximum survival rate estimate (ϕ_{max}) of 0.486 (SE 0.024), and a rate of decline (b) of 0.307 (SE 0.029).

The most important problem to be aware of in any of these approaches is bias resulting from observation error. To understand this bias, imagine a series of abundance estimates taken from an essentially density-independent population. Whenever census error leads to a mistakenly high observed abundance, the following growth rate estimate will be biased low, leading to a spurious negative correlation between abundance and growth rate and general over-estimation of the strength of density dependence. The same principle applies to estimates of density dependence in survival or productivity. This bias can be corrected if the degree of observation error is known, either through formal statistical methods based on sampling or, if no sampling has been used, through repeated surveys of the same population to measure the error directly. Alternatively, promising simulation methods that may help are now becoming available. Freckleton *et al.* (2006) provide a detailed review of this problem and how to deal with it.

age or size classes, as younger classes tend to be more susceptible to competition, and may be the key point of population regulation (Eberhardt 2002).

There are important pitfalls that you should be aware of in attempting to quantify density dependence. Although density dependence is a universal property of natural populations, it is surprisingly difficult to measure, and great effort is usually required to do so. Precisely because of density-dependent regulation, natural populations may fluctuate rather little, giving us very little information about the true form of density dependence. Even given data from a good range of abundances, any density responses may be obscured by the population's response to environmental variation and by observation error. The effect of population density and the environment on an individual's vital rates (their probability of survival, growth and reproduction) can depend on many factors, such as their body size, genetics and age. There is a growing body of research using high quality, long term individual-based datasets to tease these influences apart (e.g. Pelletier *et al.* 2007). These studies show that time-series of population size alone are likely to provide very **estimation** of the form and strength of density dependence. Finally, observation error is also an important source of bias, tending to result in an over-estimation of the strength of density dependence, and the correction of this tendency may require further sampling effort (Box 2.15). These caveats are not meant to put you off attempting to quantify density dependence entirely; however, it is important to be aware of them in order to judge what can reasonably be achieved with a given level of effort. An alternative approach for species which are likely to show non-linear density dependence is to use a range of more realistic functional forms, chosen based on literature about similar species (Box 2.15), and then carry out sensitivity analyses to see what effect your assumptions have on your results (Sections 4.4.1 and 5.3.6).

2.4.5 Physical growth and size class transition rates

The physical size of an organism is often an important determinant of its contribution to population growth. Equally, from a harvester's point of view, size can be an important determinant of profitability. It may therefore be useful to quantify individual **growth patterns** in exploited populations in order to understand both harvester behaviour and population responses to harvest. For example, given an age-structured population model of the kind described in Chapter 5, if we know how individuals grow with age, we can predict total harvest biomass, which may be a more sensitive determinant of harvester behaviour than numbers of individuals. Box 2.16 illustrates a commonly used model for quantifying **age–size relationships**.

In some species, particularly those that are long-lived and have continuous growth potential (for example, many plants, fish and reptiles), it will usually be much easier to measure size than age. In this case, when it comes to modelling the harvest, it will probably be more convenient to use **size-structured**, rather than age-structured, models, and this approach should definitely be used when growth is not strictly linked to age (for example, plants may shrink as well as grow, Van der Voort and McGraw 2006). In these models, transition from one class to another is

Box 2.16 Defining growth: the von Bertalanffy growth function.

The von Bertalanffy growth function (von Bertalanffy 1957) is the most widely used model of physical growth. This relates linear measures of size (such as length or height), s, to age, a, by:

$$s = s_{max}\left(1 - e^{g(a_0 - a)}\right)$$

where s_{max} is the asymptotic maximum size, g is the rate of growth towards maximum, and a_0 shifts the curve along the age axis (so allowing size at age 0 to be greater than zero). If growth is measured by weight, w, a modification is required to take account of the scaling of weight with size (usually a cubic power function). Allowing the weight-size scaling exponent, b, to vary, a weight-based growth curve is given by:

$$w = w_{max}\left(1 - e^{g(a_0 - a)}\right)^b$$

A non-linear procedure, such as the *nls* (non-linear least squares) function in R software, can be used to fit this model to age and size or weight data for a cross section of individuals. However, it is important to have some observations near the largest and smallest possible sizes if these model parameters are to be estimated precisely.

The addax *Addax nasomaculatus* is a Saharan antelope critically endangered by hunting, with probably fewer than 200 now remaining in the wild (Wacher, *et al.* 2005; Figure 2.15). However, since the mid-1980s, a herd of 50 addax has been held in a 10 km² enclosure in Bou-Hedma National Park, Tunisia, with a view to eventually re-establishing a wild population in the northern part of the historic range. As part of an assessment of the demographic performance of this herd, its growth curve was compared with that of a fully captive herd in St Louis Zoo, USA. From the Tunisian herd, 19 addax were captured, aged from tooth eruption and wear patterns, and weighed, while ages and weights were available for 11 animals in the zoo population. Using a fixed weight exponent (b) of 2.6 (estimated from weight and chest girth data), models were fitted to the data for each population separately, and for all animals combined, yielding the following estimates (standard errors in brackets):

	w_{max}	g	a_0
Tunisia	116.9 (6.7)	0.75 (0.17)	−0.82 (0.43)
St Louis	108.3 (4.5)	2.22 (3.58)	0.19 (1.34)
Combined	111.2 (3.9)	0.85 (0.18)	−0.8 (0.41)

Fitting separate curves for each population did not provide a significantly better fit ($F_{3, 24} = 1.74$, $p = 0.19$), suggesting identical growth curves. However, lack data for very young animals from the zoo population meant that the parameters governing initial growth (g and a_0) were very poorly estimated for this population, giving little power to detect differences.

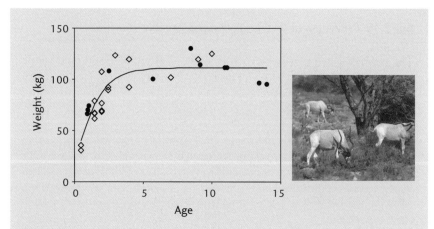

Fig. 2.15 Addax are critically endangered by hunting, so there is no suggestion that management options could aim for sustainable use in the near future, but information on growth can still be helpful in order to assess the condition of managed populations. The graph fits a von Bertalanffy growth curve to age and weight data for addax, comparing a semi-captive population in Bou-Hedma National Park, Tunisia (filled points), with addax from St Louis Zoo (open points). The Tunisian population is part of a programme to reintroduce addax to the northern part of their historic range.

Source: St. Louis Zoo, CMS/FFEM and Tunisian Direction Général des Fôrets unpublished data. Photo © Tim Wacher.

not a strict function of time as it is in age-structured models, and we therefore need to estimate these transition probabilities.

The simplest analytical approach is to use **known fates**, tagging a sample of individuals and monitoring them over time to observe directly the proportions of each class that change state in each time unit. While this works well for plants and other sessile species (e.g. Van der Voort and McGraw 2006), for most animals, less than certain re-observation from year to year will necessitate a mark-recapture approach. In this case, an extension of open population survival analysis known as **multi-strata modelling** can be used (Nichols *et al.* 1992; Williams *et al.* 2002). This can provide estimates of the rates at which individuals move from one state to another (the probability of growth from one size class to the next in this case), while controlling for survival and detection rates. The approach requires the size class of each individual to be recorded on capture or re-observation, but otherwise the same assumptions and considerations apply as those discussed in Section 2.4.2.2. A final possibility when size and age are closely correlated is to derive transition rates from a **growth curve and rates of survival** (Box 2.17), although this is a data-hungry approach, requiring both age-specific survival rates and estimates of age-specific size in order to plot the growth curve.

Box 2.17 Estimating size class transition rates.

The probability that an individual in a given size class will grow to the next one in a given time unit is equal to the proportion of individuals in the class that are within one time unit of the size threshold for transition. If it can be assumed that survival rate, S, is constant across all members of the class, the transition probability for a class that lasts t time units is given by:

$$q = \frac{S^{t-1} - S^t}{1 - S^t}$$

The average time span of a size class, t, can be measured directly by following the growth of a sample of individuals. However, it will obviously take a very long time to accumulate the necessary information in this way for long-lived, slow-growing species. In this case, data on the sizes and ages of a cross section of individuals can be used to plot a growth curve, and the average time span of a size class can be estimated from this. For example, given growth rate (g) and maximum size (s_{max}) estimates from a von Bertalanffy growth function (see Box 2.16), the average duration of a class with upper and lower size thresholds of s_u and s_l is given by:

$$t = \ln\left(\frac{s_{max} - s_l}{s_{max} - s_u}\right)\frac{1}{g}$$

2.5 Spatial issues

Harvesters and the species they harvest exist in a spatially variable world, and analyses of sustainability will often need to take this into account. This is particularly the case where spatial management strategies, such as zoned use areas, have been implemented or are being considered (Section 6.4.1). The main analytical approaches that focus on spatial patterns, and which we introduce in this section, are the estimation of **movement rates**, and the measurement of **habitat associations**.

2.5.1 Movement rates

In mobile species (including plants with widely dispersed seeds or corals with pelagic larvae), movements in or out of a population can be an important element of its dynamics. For example, if regular immigration boosts the growth rate of an exploited population, the sustainability of its use might depend on maintaining the source of the immigrants. Indeed, this process is a key determinant of the success or otherwise of no take zones or protected areas in preserving stocks and providing overspill for harvest (Sale *et al.* 2005).

Direct measures of movement depend on observing the locations of a sample of **known individuals** at intervals. One way to do this is by **telemetry**. Traditionally this involves attaching tags that give a radio signal, which is then located by triangulation with a hand-held or vehicle-mounted receiver. However, wildlife telemetry is being revolutionised by technological advances, with options now available for automated location recording and data recovery based on global positioning systems (GPS), the Argos satellite system, cell phone networks, and static hydrophone tracking of acoustic tags in the marine environment. Alternatively, animals can be **marked with tags** that can be identified when they are either recaptured, recovered dead, or observed in the field. Leg rings on birds (either metal or coloured plastic to enable reading in the field) are the classic example of this kind of marking, but technology has again expanded the range of species and situations to which marks can be applied, for example, subcutaneous passive internal transponder (PIT) tags, visible implant fluorescent (VIF) tags, coded wire (CW) tags, and visible implant elastomer (VIE) tags can all be applied to smaller species that may otherwise be difficult to mark. Section 2.7 provides published and web resources on all these forms of tracking. While telemetry clearly gives a much better chance of relocating marked animals, visible tags are cheaper, and many more of them can usually be fitted. In some cases, visible tags may thus increase the chances of picking up relatively infrequent, but ecologically significant, long-distance movements.

Movement rates based on periodic observations at two or more locations are best analysed by using **multi-strata mark-recapture models**, which separates the probability of movement from the probabilities of survival and detection (Lebreton *et al.* 2003). Box 2.18 gives an example of this method in action. Alternatively, **combined live-dead analysis** can give an indication of the rate of emigration from the area in which live sightings are recorded if the dead recoveries come from outside this area (Frederiksen and Bregnballe 2000).

Box 2.18 Migration analysis in action: quantifying Canada goose movements.

The Atlantic Canada goose *Branta canadensis* population has a wide distribution in winter, spanning much of east coast USA (Figure 2.16). While the population as a whole fared well over the second half of the 20th century, different sections of this population showed strikingly different patterns of change, with the more southerly part of the range showing a sustained decline. Hestbeck *et al.* (1991) analysed the movements and survival of the wintering population along the east coast of the USA in order to determine whether the changes in sub-populations were due primarily to differences in survival rates (potentially linked to harvest rates), or simply to a redistribution of the population.

The data used come from almost 29,000 geese fitted with coded neck collars over three years between 1983 and 1986, and over 100,000 re-sightings of these geese were made up until 1988, assigning all records to one of three locations: mid-Atlantic in the north, Chesapeake in the centre of the range, and the Carolinas in the south. Using a multi-strata capture recapture model, controlling for survival and sighting probabilities, the probabilities of moving from one sub-population to another were estimated as:

	Year $t+1$		
Year t	Mid-Atlantic	Chesapeake	Carolinas
Mid-Atlantic	0.71	0.287	0.003
Chesapeake	0.096	0.889	0.015
Carolinas	0.067	0.371	0.562

The low degree of fidelity to the Carolinas sub-population provides some support for redistribution of the population away from the south as an explanation for decline in the Carolinas, perhaps driven by improving conditions in the north allowing geese to shorten their migration (a phenomenon observed in several other migratory goose species). However, the high rates of fidelity and immigration to the Chesapeake region do not match the decline observed there during the study period. Differences in survival rates were a more likely explanation for changes in the more northerly sub-populations.

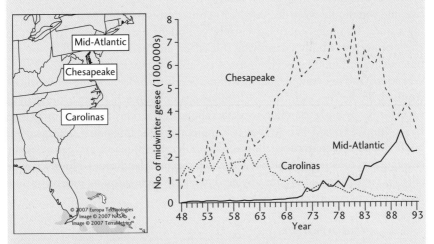

Fig. 2.16 Atlantic Canada geese have increasingly been adjusting their migration behaviour to spend the winter further north (a process known as short-stopping). This is in response to trends in food availability, driven by a combination of changing farming practices and climate warming.

Source: Hestbeck *et al.* (1991).

Measuring movement when individuals cannot be tracked is more problematic. One possibility is to analyse the **genetic structure** of sub-populations and use a genetic model to estimate recent rates of gene flow, and hence migration (Wilson and Rannala 2003). Another possibility is to fit a **population model** incorporating migration to catch and abundance data, and test whether this fits the data better than a model without migration (Section 5.3.6). Finally, for propagules dispersed by wind, current or animals, independent experiments might be carried out to quantify the pattern of **settlement**. For example, for wind dispersed plants, an array of seed traps (containers or sticky surfaces) can be placed at varying distances from focal plants (e.g. Dauer *et al.* 2007; Gomez-Aparicio *et al.* 2007), while for animal-dispersed seeds, behavioural studies of the key dispersing species can be used to quantify movements (e.g. Weir and Corlett 2007).

2.5.2 Abundance–environment relationships

Understanding the determinants of spatial variation in abundance can enhance our ability to assess sustainability. Whenever an abundance survey is carried out by sampling at a number of sites, it is therefore worth considering the possibility of measuring some key environmental variables at each site. In designing a survey of this kind, there are number of points to consider.

Survey aims

Studies of abundance in space fall broadly into two types: correlative and predictive, and you should try to decide which approach you are aiming for when designing a study. In a purely correlative study, you might simply attempt to detect an effect of harvest, ideally quantifying this effect, and perhaps controlling for any potentially confounding environmental variables. If the impact is expected to be very strong and not complicated by habitat, relatively little sampling effort will be required. Predictive studies use essentially the same approach, but take the analysis a step further by using the results to predict abundance at times and places other than those sampled (Box 2.19). As well as its potential value in guiding management, predictive power can improve the precision of density estimates, and can allow the estimation of total population size across heterogenous landscapes. However, to be reliable, predictive studies generally need to ensure high sampling intensity across several key environmental variables.

What to measure

Large-scale surveys are highly labour intensive, and may only be possible if the effort required to estimate abundance at each site is minimised. In practice, this often means a **single sample plot** at each site, such as a transect or point count, yielding either presence–absence data, or an index of abundance in the form of raw counts of individuals or signs seen. Analysis of indices of this kind can be revealing, but it is essential to remember that these indices may reflect variation in detection probabilities as much as true abundance. An appropriate census method should ideally be used in at least a sub-sample of sites in order to test whether there is significant variation in detection probabilities between them (Section 2.3). For example, in the case of presence–absence data, **occupancy models** can be used to quantify variation in detectability with site characteristics (Royle, Nichols and Kéry 2005, Section 2.3.6.2). Failure to account for detectability may seriously undermine your study. For example, where hunting causes target species to become more wary of people, uncritical use of an abundance index based on sightings will overestimate the impact of hunting.

The choice of environmental variables to measure as possible **correlates of abundance** is often huge, and some educated judgement is required in order to identify a limited set of candidate variables that are both measurable and likely to have an important influence. Ideally, a direct measure of **harvest effort** such as the total number of harvesters active per unit area will be used, although proxies such as distances to roads or settlements might be used if there is strong reason to believe that they correlate well with harvest effort. **Habitat variables** might be broad classifications of type, or focus more closely on aspects of the ecology of the focal species such as the occurrence or availability of key resources and constraints. For example, in the case of plants, soil type and moisture might be candidate variables, while for an animal, one might look for factors related to the availability of food and breeding sites.

Environmental variables can be measured **on the ground**, e.g. by visually classifying habitats at survey sites, or by measuring quantifiable factors such as soil fertility. In addition, many variables can be measured by **remote sensing**. For example, altitude, distances to roads or settlements, and coverage by water might be taken from topographic maps, while habitat type might be measured using aerial photographs or remotely sensed satellite imagery. Spatial data of these kinds are best manipulated in geographical information systems (GIS) in order to calculate summary statistics such as the density of roads in each survey site. The benefit of measuring potential correlates of abundance remotely is that they can be measured across entire landscapes. Assuming that a model with good predictive power is obtained from the sample of survey sites, this allows the focal species' abundance across the landscape to be predicted without the need for further ground surveys.

Definition of sampling units

The results of spatial analysis are often sensitive to the **size of sampling unit** (Boyce 2006). For example, for a small plant whose abundance is largely determined by local conditions on the scale of metres, variables measured at the scale of kilometres will not pick up the relevant factors. Conversely, for a highly mobile animal, fine-scale measures may fail to pick out any responses, or reflect small-scale movements rather than large-scale habitat suitability. For the purposes of assessing harvest impacts, the scale at which harvesters operate will also be relevant. Small-scale harvest on foot will vary at the scale of one to a few kilometres, while large-scale motorised harvesters may operate over tens of kilometres or more, and sampling units should be defined accordingly. Where the ecology of the species operates on a different scale to that of the harvest, the finer scale should be used to define sampling units. The scale of measurement should be similar for abundance and predictor variables.

Survey design

In order to ensure that gradients in important environmental variables are fully covered, the ideal survey design is to **stratify** by key variables (Hirzel and Guisan 2002; Vaughan and Ormerod 2003). To do this, identify which habitats or influences are likely to be most important (harvesting effort will be a key one), categorise these variables across their ranges, and divide the study area into regions according to combinations of these categories. Equal sampling effort can then be allocated to each region. Where the focal species is rare or localised, it may be necessary to increase sampling effort in sites known to have good abundance (effectively stratifying by abundance). However, the analysis and interpretation of data collected in this way is more complicated (Keating and Cherry 2004) and the approach is not analytically ideal.

Sample size

The **number of survey sites** required to deliver robust measures of association depends primarily on the number of candidate explanatory variables. A basic correlative study of harvest impact may be possible with just a handful of sites, but ideally, if the analysis is to be used predictively, at least ten sites should be covered for each environmental variable considered.

Analysis and interpretation

The most accessible approach to the analysis of spatial patterns of abundance is to use a **generalised linear model** to identify significant predictors of species' abundance (for example see Boxes 2.19 and 4.4). This approach is flexible (methods are available to deal with either binary presence–absence data, raw counts or absolute abundances), and readily available in standard statistical software (Section 2.7). However, there are other approaches which might be considered, for example in order to account for spatial auto-correlation in your data, or in order to cope with large numbers of candidate predictor variables. Guisan and Thuiller (2005) provide a useful review of recent developments and current tools in this area.

If spatial analyses are to be used predictively, their reliability should ideally be **validated** by comparing predicted abundance or occupancy with the actual situation in sites other than those used for the original analysis (Vaughan and Ormerod 2005). This requires further sampling, which should be taken into account when forecasting the sampling effort required. In the end, though, it is often difficult to identify good spatial predictors of abundance, even with a great deal of effort. Constraints on sampling effort, failure to identify good candidate variables, and the difficulty of measuring key variables at the right scale are common reasons why the attempt might come to nothing. It is also important to be aware that even where significant associations are identified, highest densities do not necessarily indicate the **best habitats**. For example, in a population whose distribution is not at equilibrium, low density or absence may simply reflect a failure to have fully colonised all suitable habitat. Similarly, behavioural responses to human disturbance can alter species' habitat use away from otherwise preferred habitats.

Box 2.19 Spatial modelling of green peafowl *Pavo muticus* distribution.

Green peafowl are native to the forests of south-east Asia, where they are declining rapidly due to hunting and habitat loss (Figure 2.17). Brickle (2002) set out to identify the determinants of the species' abundance, and thereby predict its distribution across Dak Lak province, Vietnam. Survey teams visited 161 points, selected to represent the major habitats present, but with selection logistically constrained to be near road access. Points were visited during the peak of male calling activity, and the minimum number of breeding males at each point was identified by locating unique calling individuals during two-hour visits (calls can be heard up to a kilometer away). A range of environmental variables were measured around each site, including broad habitat type and distances to the nearest permanent water bodies and human settlements. Using a generalised linear model with Poisson error structure, these three variables were found to be significant predictors of the minimum number of males present at a site. Habitat cover and the distance variables were then measured across the whole province using a GIS database derived from a combination of satellite imagery and topographic maps. Combining GIS data with the results of the generalised linear model, the abundance of peafowl could then be predicted across the province.

Green Peafowl were found only in forest, predominantly deciduous. Abundance in the next most suitable habitat (mixed forest) was half that in deciduous forest, and land within hunting distance of human settlements had average peafowl abundance less than half that in less accessible regions. While the reduction in abundance near settlements may be partly due to disturbance and habitat alteration, hunting is likely to have a strong impact, and the degree of reduction in abundance suggests that this hunting is at least locally unsustainable. The most suitable habitat free of hunting was predicted to support 72% of the total peafowl population of Dak Lak in only 17% of the land area, indicating that the population is extremely vulnerable to increased human access to forest. The study used a relatively crude index of abundance, and may have biased sampling due to the use of roads for access. However, in this case, these potential problems are the price paid to facilitate a large, widespread sample of points, and it can be argued that any biases are likely to be minimal. The result is a useful indication of the current severity and spatial distribution of threats to green peafowl in Dak Lak province.

Fig. 2.17 Based on call surveys across Dak Lak Province, Vietnam, green peafowl are most abundant in the northwest of the province, especially near rivers away from human settlement. Darker shades on the map represent greater relative abundance, assuming that call frequency correlates well with abundance. *Source:* Brickle (2002). Photo: © Nguyen Tran Vy.

2.6 Surveying as a component of monitoring

This chapter shows how it is possible to gather detailed information on many aspects of a population's demography, but you of course also need to consider **practical issues**. Surveying wild species is generally costly, in terms of both time and money, and some-times personal risk as well. For example, mark-recapture methods usually require very substantial effort to get good results, perhaps requiring a lot of manpower to deploy large numbers of costly traps over sufficiently large spatial scales, and there are often risks of injury or disease in handling animals, both to the handler and the handled. Direct observation methods such as distance sampling can be more efficient, but not always—costly equipment may be needed (such as laser range finders for measuring distances), and scarce or inconspicuous species will require huge distances to be covered to yield enough data for a meaningful assessment. Weighing up these practicalities requires a good understanding of the **field methods** that are applicable to your focal species. Sutherland (1996) provides a useful overview of the survey methods that are commonly applied to different groups of plant and animal, and there are many other inventive methods out there—use web and literature searches to find out what other people have done with species similar to yours.

The costs of surveying need to be weighed against the potential **benefits**. Some of the sustainability assessment methods in Chapter 4 require no direct biological information, and this approach may be adequate if you are explicitly not interested in the biological component of sustainability. Otherwise, at least some biological information will be required, usually abundance as a minimum in order to give a picture of the current state of the population (Chapter 4). A more detailed understanding of the processes that lead to change in population size can be gained by measuring demographic rates such as survival and productivity, and this understanding can be used to turn monitoring information into a predictive tool (Chapter 5). The trick is to decide what information is essential, and what is not justified by the expected degree of increase in your understanding.

So, before deciding on a survey strategy for an exploited species, define a range of questions that could be tackled, assess the time and financial costs of the methods necessary to address them and weigh these costs against the expected benefits. For example, if your species is productive and fast-growing, an age-structured model is unlikely to give you much of an improvement in predictive power over a simple unstructured model (see Section 5.4), so the extra effort required to gather age-specific demographic rates in the field will almost certainly not be justified. By identifying the most efficient approach in this way, you can avoid wasting time and resources attempting to gain data at a level of detail that turns out to be impossible or unnecessary. We expand on the theme of monitoring and decision making under information and resource constraints in Section 7.4.

2.7 Resources

2.7.1 Websites

Software

DISTANCE. Distance sampled line and point transects, including spatial analysis and mapping: http://www.ruwpa.st-and.ac.uk/distance/index.html

DENSITY. Spatial Capture-Mark-Recapture analysis: http://www.landcareresearch.co.nz/services/software/density/index.asp

MARK. Huge suite of mark-recapture analyses for survival, abundance and more: http://www.warnercnr.colostate.edu/~gwhite/mark/mark.htm

CAPTURE. Useful web-based analysis engine for closed population mark-recapture abundance estimation: http://www.mbr-pwrc.usgs.gov/software.html

PRESENCE. The models in Royle *et al.* (2003, 2005): http://www.mbr-pwrc.usgs.gov/software/doc/presence/presence.html

BAND2. Sample size for dead recovery analysis: http://www.mbr-pwrc.usgs.gov/software.html

R. A powerful and flexible programme for implementing virtually any statistical analysis, including randomisation and bootstrapping, known fate survival analysis and GLMs for habitat analysis: http://www.r-project.org/

USER (User Specified Estimation Routine). Can be used to fit multinomial models for change in ratios, productivity and survival based on counts: http://www.cbr.washington.edu/paramest/user/

MULTIFAN-CL. Implements statistical catch-at-age models based on either length or age structure of offtake: http://www.multifan-cl.org/

RANGES Analyses spatial data, including home-ranges, dispersal distances and habitat associations. Not free: http://www.anatrack.com/

RESAMPLING. A basic package for simple randomization and bootstrapping procedures, including a useful introduction to the subject: http://www.uvm.edu/~dhowell/StatPages/Resampling/Resampling.html

Other

Patuxent wildlife software page, a mine of useful packages: http://www.mbr-pwrc. usgs.gov/ software.html

Sampling theory tutorial: http://stattrek.com/

NASA remote sensing tutorial: http://rst.gsfc.nasa.gov/

ESRI's GIS guide (with links to books, etc.) and conservation page (grants, courses, software): http://www.gis.com/, http://www.conservationgis.org/

Radio, satellite and GPS tracking products and information: http://www. atstrack.com, http://www.biotrack.co.uk/, http://www.anatrack.com/

StatSoft online tutorial on time-to-event survival analysis: http://www.statsoft. com/textbook/ stsurvan.html

2.7.2 Textbooks

Buckland, S.T., Anderson, D.R., Burnham, K.P., Laake, J.L., Borchers, D.L., and Thomas, L. (2004). *Introduction to Distance Sampling: Estimating Abundance of Biological Populations*. Oxford University Press, Oxford. Comprehensive coverage of the theory and application of this important method. There is an online version of the 1993 edition available for free download at http://www.colostate.edu/Dept/coopunit/download.html.

Hilborn, R., and Walters, C.J. (1992). *Quantitative Fisheries Stock Assessment: Choice Dynamics and Uncertainty*. Kluwer, Dordrecht. Essential reference and thought-provoking discussion on how it's done in the commercial fisheries world.

Kenward, R.E. (2001). *A Manual for Wildlife Radio Tagging*. Academic Press, San Diego, CA. A useful summary of the practical and theoretical issues in radio tracking.

Marzluff, J., and Millspaugh, J.J., eds. (2001). *Radio Tracking and Animal Populations*. Academic Press, San Diego, CA. An edited volume of articles surveying the state of the art in wildlife radio tracking.

Skalski, J.R., Ryding, K., and Millspaugh, J.J. (2005). *Wildlife Demography: Analysis of Sex, Age and Count Data*. Elsevier Academic Press, Amsterdam. A technical tool box of quantitative methods with an emphasis on those commonly applied to North American game animals.

Sutherland, W.J., ed. (1996). *Ecological Census Techniques*. Cambridge University Press, Cambridge. A handy overview of field methods, arranged by taxonomic group, including the measurement of environmental variables.

Thompson, S.K. (2002). *Sampling*. John Wiley and Sons: New York, NY. Comprehensive reference on the statistical background to getting a decent sample.

Williams, B.K., Nichols, J.D., and Conroy, M.J. (2002). *Analysis and Management of Animal Populations*. Academic Press, San Diego, CA. Hugely detailed coverage of modern methods for abundance and demographic analysis, also covering population dynamics and population management issues.

3

Understanding natural resource users' incentives

3.1 Scope of the chapter

In this chapter, we will discuss some of the general issues involved in collecting data about people's motivations for natural resource use. We cover the main techniques for data collection and analysis, and then look at some case studies which illustrate the points made in the chapter.

The relationship between humans and their natural environment is studied in a huge range of disciplines, spanning the arts, natural and social sciences. These include history, politics, theology, cultural studies, development studies, ethics, economics, archaeology, ecology and anthropology. For example, the pattern of trade in wild pigs in North Sulawesi is determined by the distribution of pork taboos—only the Christian populations at the tip of the island are interested in eating wild pig meat (Clayton *et al.* 1997). In many parts of the world, including China and Russia, areas of natural forest are preserved not for resource use but as sacred groves (Laird 1999). Interpretations of the way in which humans interact with their environment range from the extremes of the romanticised 'noble savage' (Redford 1990) to the rational exploiter who destroys for short-term gain in the 'tragedy of the commons' (Hardin 1968).

In this book we approach the issue of assessing sustainability primarily from a natural science perspective, with an emphasis on hypothesis-testing using quantitative data. However, when researching people's behaviour towards natural resources, we need to step into other disciplines. The scientific way of thinking is also used in some areas of social science, particularly economics. But other research philosophies emphasise the importance of starting off by defining the theoretical construct within which you are working, or focus on understanding and interpretation rather than explicit hypothesis-testing (Strauss and Corbin 1998; Mehta *et al.* 1999). These differences can lead to misunderstandings and disagreements, but also allow interdisciplinary researchers a much richer understanding of the problems that they are studying. So it is worth remaining open to information and ideas from as wide a range of disciplines as possible.

3.2 General issues

3.2.1 What do we need this information for?

Given the wide range of approaches available when researching people's relationships with their natural resources, it is especially important to define the research question precisely in advance. There are far too many aspects to the problem of over-exploitation for one study to attempt to address them all, and the disciplinary framework that we use depends to a great extent on the questions to be answered (Table 3.1).

One major issue is whether the study will be used for scientific inference, or whether it rather requires the researcher to develop a deep qualitative understanding of the issues. The conceptual divide between **qualitative and quantitative research** is quite sharp in social science. So, for example, a study into the relationship between poverty and natural resource use might choose to collect quantitative data on income levels for a randomised sample of people, and correlate this with their resource use patterns. Alternatively, it could focus on particular individuals who have been through crises of extreme poverty, and aim to understand how they adjusted their livelihood activities to cope with the situation. In practice, the two approaches are not exclusive. A combination of approaches enables researchers to gain insight from the strengths of each (Carvalho and White 1997), and often it is useful to start with qualitative research as a way of informing a subsequent quantitative study (Section 3.2.4.1). This can help you to understand the wider context of the issue before focusing on the specific questions of interest to you.

A second determinant of the best methods to use is the extent to which the study is linked to **conservation action** rather than being pure research. Data-gathering in conservation projects often has a dual agenda so that while researching the situation, the researcher is also trying to raise public awareness about the issues and to engage stakeholders in conservation. In this case, the tools for data collection are likely to be more participatory and less focused on collecting quantitative data. This dual agenda is often unspoken, but it needs to be considered, because there is a danger that the neutrality and objectivity of data collection for research purposes can be tainted by researchers approaching the issue with an underlying agenda for action. Hence it is usually advisable to separate research and monitoring from conservation action as far as possible. Then the data collected can provide independent feedback on the success of the action. For example, market monitoring is best carried out by people unconnected to any conservation programme. Note that this advice is contrary to the philosophy of much development research, which has a strong participatory element such that research and action become intertwined (Chambers 1992, see below). Which approach is best to take is case-specific, but the decision needs to be explicit and well justified.

Third, there are some key divisions in the types of data that can be collected, which influence the uses to which the data can be put. The first is between **perceptions and actions**. Any change in the behaviour of users towards their natural resources has three components: their attitude changes; they act in a different way;

Table 3.1 Broad classification of the questions that can be asked about natural resource users, and the techniques that can be used to answer these questions.

Note: In this chapter we concentrate on the first part of the list. This is because our focus in this chapter is on data collection at the scale of the individual or household. Issues of resource management and the interaction between policy and sustainability are dealt with in Chapters 6 and 7. The case studies examined in detail later in this chapter are given in bold.

Question	Scale	Framework	Factors	Techniques	Examples
What factors influence the decision to use natural resources	Individual–Household	Cost–benefit analysis	Direct and indirect costs of hunting/foraging (including opportunity costs); revenues obtained, risk of capture if illegal	Interviews, user and ranger follows, monetary valuation	Milner-Gulland and Leader-Williams (1992), **Abbott and Mace (1999)**
How do users choose species and locations	Individual	Optimal foraging theory	Encounter rates, offtakes, wastage levels; catch per unit effort	Follows, interviews, focus groups	Rowcliffe *et al.* (2003), Alvard (1993), **Muchaal and Ngandjui (1999)**
How do socio-economic characteristics affect the type and level of offtake	Individual–Community	Sustainable livelihoods, correlational studies	Distance to markets, resource tenure, livelihood options, alternative foodstuffs, age	Interviews, direct observation	Cinner and McClanahan (2006)
How does natural resource use fit within a household's livelihood activities	Household	Sustainable livelihoods	Seasonal calendars, food entitlements, time-budgets, wealth indicators	Interviews, direct observation, rankings, focus groups	DFID (2001), **de Merode *et al.* (2004)**
Is natural resource use important to people's food security	Household	Food security	Coping strategies, frequency of use, seasonal calendars	Rankings, interviews, focus groups	Maxwell and Frankenberger (1992)

Table 3.1 (Con't.)

Question	Scale	Framework	Factors	Techniques	Examples
What are people's perceptions of the roles of natural resources and/or conservation	Individual– Community	Qualitative analysis	Natural resource use as a cultural activity, perceptions of risk, relationships with authorities	Attitudinal surveys, focus groups	Holmes (2003)
How has natural resource use changed over time	Individual– Community	Qualitative analysis	Changes in effort, gear types, prey density	Key informant time-lines, case studies, historical literature	Freehling and Marks (1998)
What determines consumption of natural resources	Household	Consumer choice	Multivariate analysis of factors affecting consumption (e.g. price, availability, taste)	Household interviews, recall of consumption, focus groups, rankings	**Fa *et al*. (2002), Wilkie and Godoy (2001)**, East *et al.* (2005)
What determines consumption of natural resources	Market	Econometrics	Time-series of quantities on sale, prices (the good itself and substitutes), incomes, inflation rates	Direct observation, official statistics	Milner-Gulland (1993), Wilkie *et al.* (2005)
How do natural resources flow from the hunter/gatherer to the consumer	Individual– Market	Commodity chain analysis	Estimating profits along the chain, quantifying flows between actors, identifying targets for intervention	Direct observation, key informants	Cowlishaw *et al.* (2005)
What is the total economic value of natural resource use to individuals and society	Individual– Society	Environmental economics	Monetary valuation of benefits and costs to stakeholders	Willingness to pay questionnaires, substitute goods, direct valuation	Bann (1998), de Lopez (2003)
How do institutions and policies affect natural resource use behaviour	Institutions– Individual	Institutional analysis	Studies in sites with contrasting institutional regimes, flows of influence between stakeholders	Key informant interviews, focus groups	Alcala (1998), Ostrom *et al.* (1999), Béné (2003)

and they perceive that they have changed their behaviour. The reported change in perceived behaviour is not necessarily the same as actual change in behaviour. This may be due to people not telling the truth, but often it is at least partly because what people think they are doing isn't what they are actually doing. So, for example, if we wished to assess whether conservation action had led to people hunting an endangered species in a more sustainable way, we could ask how people now feel about the species, or whether they have changed their hunting behaviour, or we could observe whether hunting mortality has indeed declined. Only the last of these is directly related to extinction risk (although even then there is still the question of whether reduced hunting mortality is actually leading to species recovery; the methods in Chapter 2 are needed to shed light on this). Holmes (2003) examined 18 studies of community attitudes towards conservation, and found only 2 that considered whether attitudinal changes had indeed led to behavioural changes. The others either assumed, or stated explicitly, that attitudes were an adequate proxy for behaviour. It is fine to measure community attitudes to conservation if that is your research question, but it is important to be aware that attitudes, perceptions and action are not equivalent.

Studies of sustainability concern change over time. However, the past cannot be measured directly. If there are no contemporary data available, then inevitably we need to rely on people's **recall of the past**. Hence we might ask questions about how the abundance of a species has changed over the last 10 years. However, the past is filtered through people's perceptions, whether it is 1 day or 30 years ago. Perceptions may or may not be an accurate reflection of actual events. There are techniques that can help minimise biases when using recall data, which are discussed below. But it is worth thinking in advance about whether it is possible to get data that are not based on recall.

Finally, human relationships with the environment take place on **many scales**. At the smallest scale it is individuals who kill or collect animals or gather plants. But they live within a social network, including the household, the immediate community, trade networks, and the larger-scale institutions, such as national and local government and international treaties. Conservation action can occur at any of these scales, and is often characterised as being 'top-down' (targeting governments and other large-scale institutions) or 'bottom-up' (targeting individuals and communities). Although this characterisation is simplistic, it is important to consider early on what the appropriate scale of investigation is for your research question.

3.2.2 Ethical issues

Research on the sustainability of wildlife use is an ethical minefield. It is particularly so because unsustainable wildlife use is often illegal. There are a number of codes of conduct that can help you to follow best practice (see the end of the chapter for links). Here are some particularly relevant points:

- It is important to avoid creating **unrealistic expectations**. Just by being in a community carrying out research, you may lead people to expect benefits.

You need to be sure that they understand the likely outcomes of your research, and how it may affect them. This applies also in your relationship with local research collaborators and government officials.

- It is necessary to **tell the truth** about what your research question is and why you are carrying out your study. This can be extremely difficult, particularly when you are researching illegal activity. If it is impossible for you to tell the full truth, then you need to think carefully about how this might compromise your integrity, and take advice before you start (for example from an ethics committee or a professional association).

- A key tenet of social research is that the researcher should **do no harm** to the people they are studying. This is usually taken to mean that you should not upset people by your questioning. But in the case of conservation, the results of your research may lead to policy action in the pursuit of a social good that could damage the livelihoods of people you have interviewed (such as preventing ecosystem destruction or species extinction by removing people's rights to use them). If there is any question that this might be an outcome of your research, you need to think hard, and take advice on the ethical implications of what you are doing.

- **Don't do anything illegal.** For example, if hunters are working illicitly inside a National Park, you should be sure that going out on hunting or gathering expeditions with them is legal before you do it. Think carefully about the potential consequences of ignoring this guideline.

- Don't do anything that may **alter people's behaviour**. For example, paying a hunter per animal that they allow you to weigh may well increase hunter effort, with negative consequences for sustainability. Similarly, buying endangered species in the market can inflate demand. It may be important to check that quoted prices are the same as purchase prices, but one way to do this would be to ask people how much they paid after they have bought a piece of meat.

- In general it is a bad idea to **pay people** other than your official research assistants for information, as this alters the relationship between respondent and researcher. Small gifts to say thank you for help or hospitality are acceptable, for example pencils and notebooks for children, or contributing refreshments when holding a discussion group.

- Let people know that their answers will be treated **confidentially**, and ensure that they are. For example, publish summary results only, and use identifying codes rather than names in databases. Tell people that they have a right to refuse to participate in your study.

- It is important that you try to **give something back** to the people you are working with, rather than simply extracting information. This may be as simple as ensuring that a summary of your research results is given to the community head. But it should also include making sure that you acknowledge inputs by others, make an effort to train local people and include them fully in your research (and subsequent publications), and leave copies of your datasets with relevant organisations. Further activities might include advising

Fig. 3.1 Researcher playing with school children, Ulan-Bel' village, Kazakhstan. Photo © Natasha Balinova.

local people about issues that concern them and helping with conservation education programmes (for example, talking about your work at local schools, Figure 3.1). There can be a fine line between good citizenship of this kind and moving away from conservation research into action.

3.2.3 Gathering social data—the basics

Although there are many approaches to data collection in the social sciences, they all encounter similar issues. Just as in the biological sciences (Chapter 2), the three main ones are sample size, representativeness and bias reduction.

3.2.3.1 Sample size

The answer to the question 'what is an adequate sample size for my study?' is not straightforward, but it is vital to the success of the research. Firstly, the answer depends on whether the approach is quantitative or qualitative. In qualitative research, sample size per se is not a useful concept. Instead the researcher is aiming to understand the system deeply, and so targets the most useful and relevant informants. Case studies are often used to illustrate the issues involved (Strauss and Corbin 1998). Time constraints mean that it is impossible to work with everyone to the same depth, and hence qualitative research may use a mixture of group and individual work to ensure that all aspects of the problem have been considered. Checks for consistency (such as comparing the responses from different groups) can indicate whether there is a need for further work.

In quantitative research, there is a more concrete tradeoff to be made. This is between the time and financial resources available to do the work and the accuracy of the results that are obtained. A particularly useful tool for assessing this tradeoff is power analysis (Cohen 1992). This involves estimating the probability that you will fail to detect an effect (for example a relationship between the price of a wildlife product and hunter effort) when it in fact exists. The power of a study depends on the sample size, the effect size and the value chosen as the threshold for significance (alpha, α). Alpha measures the probability that you reject your null hypothesis when in fact it is true (i.e. detecting a relationship when there is none), and is often set at 5%. Hence you can improve the power of your study in three ways:

- by increasing your sample size,
- by increasing your alpha (e.g. to 0.1 rather than 0.05), which means you will get more false positive results,
- by increasing the size of the effect that you wish to detect.

In general, strong relationships between variables will be picked up even if there is a lot of noise, but weak ones will come through only with large sample sizes or if error is reduced. This can be done by stratifying the sample into more homogeneous groups or including covariates. Wealth ranking can be a very useful stratification. For example, if people of similar wealth act in similar ways (for example in their meat consumption decisions), statistical tests that divide people into wealth groups are likely to have more power than tests on the whole population.

Power analysis techniques range from very simple formulae to complex procedures. There are a number of software packages available to help you with this (see Section 3.4 for details).

3.2.3.2 *Randomness and representativeness*

This is another area in which qualitative and quantitative research differ dramatically. In quantitative research, the key issue is to ensure that the sample is representative of the population as a whole. The best way to do this is to use stratified random sampling. Stratification (into groups of similar datapoints, for example by community, by wealth class) reduces the amount of unexplained error, and hence increases power. Randomisation within the strata guards against bias. Another method of sampling is systematic (for example interview at every 10th house). This is usually much easier logistically than randomisation, and hence is often used. However, it carries the danger of producing a biased sample, if there is variation in the population that is correlated with your sampling. For example, if you interview hunters every seventh day, you might always get them on market day, or always get them on their day off, and hence your results will not be representative of their weekly activities.

In qualitative research, the focus is not on randomisation to ensure representativeness. Instead, representativeness is ensured based on an understanding of the system, which the researcher builds up during the study. The researcher may actively target extreme cases, or cases that illustrate borderline situations.

There are many textbooks available on designing studies in the social sciences (e.g. de Vaus 2002, which focuses on quantitative surveys, or Patton 1990, which focuses on qualitative research).

3.2.3.3 Bias reduction

The interpretation of quantitative research results requires an understanding of two factors: the precision of the results and their bias (Section 2.2). A precise study will have low sampling error, and so have tight confidence intervals around the best estimate of the parameter value. An unbiased result will give parameter estimates that are not consistently higher or lower than the true value. So it is possible to have a very accurate estimate that is at the same time biased, or an inaccurate but unbiased estimate (see Figure 2.1). Generally, using statistically valid survey design methods, particularly randomisation, will guard against bias. Hence although confidence intervals may be very wide if there is a lot of sampling error in the study, the estimate of the mean is unbiased.

Bias is an insidious problem because it is very hard to quantify, and hence it is difficult to correct for. However, it is pervasive in the kind of studies we are discussing, even after randomisation. For example, studies that rely on recall are often biased by people's differing perceptions of the past and the present (people often feel that the past was better than the present). Similarly, if you ask hunters about their typical catch rates, you are likely to get inflated estimates, because people tend to discount days on which they catch nothing. There will also be biases introduced through your relationship with the interviewee; their perceptions of your motivations and allegiances and yours of theirs. There are several methods that can be used to reduce bias, or at least check for it. These include:

- Minimising the time-recall period (for example, ask about yesterday not last month, Figure 3.2).
- Asking about actual values rather than typical values (for example, ask about numbers caught yesterday, not what is usually caught in a day).
- Triangulating (for example, ask hunters about who they gave meat to, and then ask consumers from whom they received it; ask focus groups about the major points of food scarcity in the year, and then ask individuals the same question).
- Designing the study carefully (Section 3.2.4), and using trained local research assistants.
- Spending a significant amount of time in the community, so that your and your interviewees' understanding of each other's motivations is more closely aligned.

3.2.3.4 Some golden rules

Firstly, consider carefully in advance the **practicalities and realities** of fieldwork, particularly when you are working in a country and culture other than your own. This includes being fully prepared for all the logistical challenges, bureaucratic hold-ups and health and safety issues that may arise. This is not just for your own

Fig. 3.2 Food consumption recall survey, Rio Muni, Equatorial Guinea.
Photo © Nöelle Kümpel.

security but in order to avoid inconveniencing and potentially endangering your colleagues and hosts. They remain answerable for your behaviour after you have gone, and you can put them in a great deal of difficulty through careless behaviour, either in terms of health and safety or in terms of alienating or offending people. You will almost certainly also be required by your organisation to fill in a risk assessment form before you set out for your fieldwork, but if not you should do one anyway (Winser 2004).

You also need a full understanding in advance of the **political and cultural sensitivities** that may be involved in your work, and to have acted to minimise these. You will have to develop appropriate relationships not only with the people from whom you are collecting data, but also with your local colleagues, partner organisations and officials in the relevant Ministries. This takes time and sensitivity, particularly as each of these interest groups will have different expectations of your research, which may not align completely with yours. Don't underestimate the importance of the differences in culture and attitude that exist between you and others, and be sure to reflect upon your own behaviour and how you can change it to ensure that you are viewed positively. You need actively to **manage your relationships** with all the people you interact with, so that you have reasonable expectations of each other's part in the research. We come back to these themes in Section 7.2. All these issues are also present for biological studies (Chapter 2), despite there being fewer direct interactions with local people.

Coming back to the data collection itself, **pilot studies are essential**. It is always a false economy to ignore the need for pilots. Pilots involve several stages:

- Try the procedure on yourself. Answer the questions as honestly and objectively as possible, putting yourself in the respondent's shoes.
- Get people with good knowledge of the system and of research methods to validate the procedure. This should include both reviewing the methods and actively trying them out.
- Try the procedure out on a population that is as close to your study population as possible. Clearly when the actual population of respondents is small, it is not possible to do a large-scale pilot, or you will use up valuable respondents (ideally, pilots should not be carried out with the same people who will respond to the main survey). But you can either use a few people judiciously, or try the pilot in a different location. Ask the respondents to criticise the procedure as well as attempting to answer your questions. It can be useful to start with focus group discussions to scope out the questions, and then pilot the draft questionnaire with individuals.
- The number of people or group sessions you need in the pilot phase should be enough that by the end of the piloting period methodological issues are no longer being raised. For example, people are answering questions consistently. Generally this doesn't take long if you are responsive to criticisms.

Piloting is often useful in itself as a way of gaining a deeper understanding of the system, and will almost certainly lead to revision of the approaches being used. It is also a good opportunity for training research assistants.

The next golden rule is **don't impose on the respondents**. Minimise the length of time that they will have to spend helping you, and keep the procedures short and simple. Arrive at times which are convenient for them, and be aware of sensitivities about particular issues. This is not just courtesy; the results will also be better if respondents feel positive about their participation. On this theme, it is also worthwhile finding out whether other studies have been done in the same area, in which case there may be problems with respondent fatigue. People will be much less keen to answer your questions if someone else has interviewed them recently, so try to avoid duplication as much as possible.

Finally, **consider how the data will be analysed before you start**. Clarify your research questions, theoretical framework or hypotheses. It can help to do some mock analyses, either of imaginary data or of data from the pilot, to ensure that the results that are obtained will actually provide answers to the questions you are asking.

3.2.4 Techniques for data collection

There are many complementary ways in which data can be collected. It is likely that all will be used to a greater or lesser extent in any particular piece of research on the sustainability of wildlife use. Rather than giving detailed explanations on how to do each method, we explain their underlying philosophy, strengths and limitations. We give a few tips on how best to do each one, and pointers on where to go for more information. The five main techniques we discuss are participatory rural appraisal, individual questionnaires, direct observation, experimental economics and use of published data sources.

3.2.4.1 Participatory Rural Appraisal (PRA)

PRA is an immensely popular approach in social research. However, the PRA umbrella covers a wide range of practices, not all of which are in harmony with its underlying philosophy (Lucas and Cornwall 2003). PRA grew out of Rapid Rural Appraisal (RRA), which was developed as a method for getting maximum information in a short visit, rather than through long-term study (Chambers 1992). RRA involves what were then less conventional methods of data-gathering, such as interviews with key informants (e.g. people selected for their knowledge and expertise—community elders) and community mapping. PRA emphasises the participation of the respondents themselves in the research—they should drive the research questions and gather the data themselves. Hence a key component of PRA is the empowerment of the people being researched, so that the research is not driven by the agenda of the researcher, but by the concerns of the local people. PRA is concerned that the voices of all people, including the marginalised sections of the community, are heard. To emphasise this fundamental shift in how research is carried out, there has been a further shift in acronym, to Participatory Learning and Action (PLA). This carries further connotations of empowerment and capacity-building, such that the outsider is working as a facilitator, enabling the people to speak out about their needs, and act to improve their situation (Chambers 2003). The interchange between researcher and the people being researched is balanced, with learning in both directions. This is a very long way from traditional research.

Whether or not you decide to engage with the underlying agenda of PRA, it has spawned some useful methods for collecting data. Many of these can be used both for individuals and in group work.

- **Time-lines** are a sequence of events that are relevant to the history of the community, with dates attached (for example, a major flood, a change in political leadership, a forest fire, the arrival of a road). The time-line enables the researcher to talk about people's situation in the past with reference to relevant events. For example, 'How did you make your living before the road came?' might be a much more useful question than 'How did you make your living 10 years ago?' Time-lines are often constructed with community elders, who remember past events best.

- **Seasonal calendars** explain the main events in community life over the year. For example, they may show when the crops are planted and harvested and when the main hunting seasons occur. People can be asked to indicate how they divide their time between livelihood activities in each period of the year or when the key points of food insecurity occur.

- **Community mapping** gives the informant's perspective on the landscape, and so emphasises important aspects that outsiders might miss. The researcher can walk with key informants (for example, community elders, or the local children) and create a map of the community. For example, the informants may highlight the location of the best water source, or places where crop raiding by wildlife is particularly bad.

- In a **livelihood matrix**, the respondents list all the livelihoods that are potentially available to them, including the conservation-relevant ones that you are particularly interested in (for example, bushmeat hunting or working as a ranger). They then score these livelihoods along a range of dimensions, such as physical difficulty, risk, prestige, career progression, profitability, barriers to entry. These scores can potentially be aggregated into a ranked list. This is a useful exercise for discussion of the options available for alternative livelihoods projects (Section 6.3.3) and of people's perceptions of the relative desirability of the conservation-relevant activity.

- **Focus groups** are forums in which the issues under investigation are discussed by a group of people. For example, focus groups are often used to discuss the concept of wealth in the community, what the indicators of wealth might be, and how individual households can be partitioned into different wealth groups (Box 3.1). Focus groups should be made up of three to six individuals, selected with an eye to the group dynamics that you wish to promote. If they are aimed at discovering how people cope with temporary food shortages, a group made up of women (who cook the meals) would be appropriate. For discussions about hunting behaviour, men might be more appropriate. Usually more than one focus group is used, in order to triangulate results and obtain a range of perspectives.

- **Ranking** allows participants to express the relative weight they give to each of a range of options or outcomes. For example, people might be asked to rank different meats in terms of their taste, availability or price; or livelihoods in terms of their potential to earn cash, difficulty, stability of income, or usefulness in times of extreme need.

PRA emphasises the use of locally available, visual props. For example, respondents might use piles of seeds to partition their livelihood activities between seasons. Researchers could show cards with pictures of various animals on them to ask which ones are commonly seen, and whereabouts. Maps can be drawn on the ground with sticks and other objects, or in colour on paper—although care must be taken, as the concept of a map can be difficult for people unused to them.

The PRA research methodology emphasises inventiveness, and the use of all available mediums to find out about the important issues that affect a community. Its strength is that it should give a much truer picture of the issues that people face, and the results should be much less conditioned by researcher expectations. The emphasis on participation means that it is an excellent platform for stakeholder engagement. Conservation programmes that truly aim to be community-led and address the community's relationship with wildlife in a positive way would do well to start with a PRA approach.

There are some issues that need to be considered when doing PRA. These include the danger of unrepresentativeness, leading to a biased understanding of the issues. For example, biases would occur from talking just to the community leaders, or running focus groups in which certain groups, such as women or lower caste members, feel unable to speak. Sensitive issues like health problems or incomes cannot always be discussed effectively in groups. Literate, well-educated people may find the PRA approach patronising. Others may find the technique threatening and be uncomfortable about speaking openly, for example, mixed groups in strongly gender-divided societies or people in repressive political regimes. Finally, PRA is essentially a qualitative research methodology, and so it can be difficult to analyse the results and make generalisations. This means that it may be more appropriate as an initial approach to understanding the issues, before carrying out more detailed surveys of individuals based on the knowledge gained.

Box 3.1 Example of a PRA technique—wealth ranking.

Wealth ranking is useful for (1) giving insight into the complex meanings surrounding the concept of wealth for a community; (2) stratifying your sample for household surveys if it seems likely that the parameter of interest (i.e. natural resource use) is related to household wealth; (3) providing a single variable representing household wealth, which can be used in quantitative analyses.

A typical procedure would be:

- Identify three focus groups in the community (perhaps women, men and elders, to give a spread of perspectives), made up of three to four people. The numbers should be small, as wealth is a sensitive subject, which may not be good for large-scale discussion.
- Write an identification number for each household in the community and the name of the household head on separate cards. This may be usefully linked to a map.
- For each focus group, start with a discussion of the concept of wealth in the community—how does it connect with livelihood activity and household structure (widows, young families with many dependents, people with grown-up children sending remittances home?), how is it expressed (house construction, piped water, mobile phone?) and what influences it (health, alcoholism, work ethic, financial skills?). Do people recognise the concept of wealth as opposed to income, and of financial wealth as opposed to well-being? Often wealth is quite an alien concept, particularly in poor subsistence-based societies.
- Ask the focus group to divide the household cards into two piles—those known to at least two members of the group and those not known. Only rank the known households, and it is best not to ask them to rank the households of people in the group.
- Ask the group to discuss how best to divide the pile into groups of households of approximately the same wealth. Record the criteria that they decide upon. Generally it is easier for them to divide the households into wealth

groups (usually three to four groups) than to produce a fully ranked list. The number of wealth groups should be decided upon by the focus group, although it is very helpful for analysis if all the focus groups choose the same number of wealth groups.

- If there is major disagreement about the wealth rank of a household between or within focus groups, consider using a key informant (local research assistant or trusted community elder) to arbitrate for you. When households are not known in one or more of the focus groups, you can use the independent informant to arbitrate if the number of households involved is small, otherwise you will need to hold another focus group. Consider whether the need for arbitration reflects some real ambiguity about the meaning of wealth, or lack of clarity in the procedure. If you get many ambiguous results, this is a warning sign that the procedure is not working well.
- When carrying out your household surveys, record the presence or amount of assets that the focus groups have identified as key indicators of wealth rank. You can use this information to validate your results (by checking that there is indeed a correlation between the presence of assets identified as important and wealth rank) and if the correlations are strong, as a way of inferring a wealth rank for future respondents without repeating the focus groups.
- When analysing your results first check for consistency between focus groups in where they place households. You would expect that, with self-determined wealth criteria, the placing of an individual household might vary between groups, particularly at the margin. However, there should be no systematic bias and few households should shift more than one wealth group. If the results are consistent, you can have more confidence that the concept of wealth ranking has been accepted by the focus groups, and that the results are meaningful. You can then assign a wealth rank to each household, using the median rank from the three focus groups.

Wealth ranking is a useful tool, but it doesn't always work. Some communities find it extremely difficult to carry out exercises such as these because of their concept of wealth. Others find it too sensitive or difficult a subject for group discussion, particularly early on in your stay with them, and the method would need to be adjusted accordingly, for example by using key informants rather than groups and waiting to do the wealth ranking until you have built trust. It also only works in relatively small communities in which people know each other well and are able to rank the households. An alternative approach would be using key informants to discuss the concept of wealth, making a list of key indicators of wealth (for example, capital assets) and then using a checklist to record their presence in each household.

3.2.4.2 Questionnaire surveys

Questionnaire surveys are the mainstay of quantitative social science research. They cover a spectrum from using PRA-type approaches but with individuals to formal structured interviews. A common approach is to use a **semi-structured**

interview, in which the main questions are prepared in advance and asked at some point, but the conversation is allowed to move naturally through these questions, so that other relevant information is obtained.

The selection of a set of people to answer the questionnaire can be made in a statistically valid way (using random sampling), or by using **purposive sampling** methods (in which particular individuals or households are deliberately selected). Fully purposive samples cannot be analysed statistically, but might be appropriate if there are only a few people with the characteristic of interest, so random sampling is likely to miss them. Snowball sampling is another form of purposive sampling, for situations in which it is hard to get into the community of interest. The researcher interviews one person and asks to be introduced to another, and so moves through a network of contacts. In many cases, it is useful to combine random and purposive sampling methods. For example, if the aim was to discover whether non-timber forest product gatherers are poorer than other members of the community, the potential respondents might first be stratified—purposely broken down into NTFP gatherers and others—and then random households picked from each group. It is best if at all possible to carry out some form of random sampling, so that results can be generalised away from the sampled group through statistical inference.

There are many textbooks on how to design a proper survey (e.g. de Vaus 2002). There are also a number of websites where surveys are stored, so that researchers can get ideas about how best to phrase questions (see the list at the end of the chapter). In general, the more formal the questionnaire, the easier it is to analyse, but the more prone the results are to researchers' preconceptions. Closed questions (those that give respondents the option to choose between a set of predefined answers) are particularly dangerous, and need thorough piloting. Open questions that lead on to discussion are the best way of getting an understanding of the topic, but are often difficult to put into a statistical framework. Here are a few pointers and pitfalls:

- Keep it **short and simple**. The questionnaire should not take more than 20 min to administer and should be an enjoyable experience for the respondent (Figure 3.3). Pilot it thoroughly, and use focus groups and key informant interviews to ensure that you have an understanding of which questions to ask and how. Have an introductory session explaining who you are, and why you are doing the questionnaire, and that answers are entirely confidential. At the end, thank the respondent for their involvement and tell them how they can get more information (for example ask if they would like a copy of the results, and give your contact details).

- Start with **general questions** to get the session flowing, and put **sensitive questions** at the end. Ask sensitive questions in acceptable ways (for example getting people to point to the income band that they belong to is more acceptable than asking them directly what their income is). Think hard about the **order** in which you ask questions, so that people don't take cues from previous questions about what you are interested in (for example if you ask people about their views on the benefits of a

Fig. 3.3. Household interview on livelihood activities and attitudes to saiga antelope conservation, Kalmykia, Russia. Photo © Aline Kuhl.

Protected Area after having asked which of a set of animals they think are endangered, it will be fairly clear that biodiversity protection is what you are interested in).

- Ensure that the survey will give you information on your **dependent** variables (for example, how much hunting do they do?), your **explanatory** variables (for example, what other activities are they engaged in, how much income do they get from each?), and **background** variables that will be used to split the respondents into groups (for example, wealth rank, education level, age, sex, household size, village). Alternatively you should have a clear linkage established between different ways of obtaining each of these data categories (for example, you are collecting detailed offtake data from hunters belonging to the same households that you are now obtaining livelihoods data from).

- Always give people the option to say '**Don't know**', so that you can distinguish those who are truly neutral from those who don't know what they think.

Avoid:

- **Ambiguous questions.** For example, the common use of the word 'important', as in 'What is your most important livelihood activity in the wet season?' Importance has many dimensions. It could mean the activity that generates the most cash, takes up the most time, provides the most reliable income or that carries the most status. Each respondent (and each researcher) will put different interpretations on the word. Instead use words that are not ambiguous, and if you wish to look at all the dimensions of 'importance' ask about them explicitly.

- **Leading questions.** For example, 'Have wildlife numbers declined in your area?' may be what you actually want to know, but the direct question suggests that you are expecting any trends to be negative. The question also gives no time-frame, and may get you secondary opinion as well as personal observation. Instead ask 'Have you noticed any changes in wildlife numbers around the village since the road came to the area?' Then follow up with an open question 'Can you describe these changes to me?' Leading questions can be very subtle, and can also be connected to the respondents' perceptions of your research agenda. It is very hard for someone who is known to be a conservationist to ask neutrally about people's views on the conservation of wildlife. People have a tendency to tell you what they think you want to hear. Training local assistants to administer the questionnaire may help with this.

- **Jargon.** Don't ask 'What are your main *livelihood activities* in the wet season?' or 'Who are the main *stakeholders* in wildlife hunting?' or 'Could you describe the wildlife *commodity chain* to me?' You may need to break these questions down to cover all the subtleties of the definition, or think of another way to approach the question, such as using a diagram. If a questionnaire has to be translated, it is even more important to use simple, direct language, to avoid further ambiguity creeping in during translation.

- **Over-complex questions.** Don't expect people to fill in lots of huge and complicated matrices (for example, of seasonal livelihood activities). Don't make them do maths (for example, asking people to assign preference values adding up to 100 to species that they most like to eat). Over-complex questions will be filled in incorrectly or left blank, and will put people off answering further questions. On a related note, questions that are obviously designed to test knowledge are not a good idea, as people feel uncomfortable being tested. There should be no obviously 'wrong' answers.

3.2.4.3 Direct observation

There are many forms of direct observation data. These might include recording the number and price of species on sale in a market (Box 3.2), the catch that hunters bring into the village each evening, the amount of grain in each family's granary, or the number of consumer goods that a household owns. Following people as they carry out the activity of interest and recording what they actually do can be very useful, though it's labour-intensive for you, gives small sample sizes, and is potentially disruptive for them. Direct observation is an excellent complement to other data collection methods. It is particularly useful for triangulation of other data sources (for example, comparing recall data on the price paid for wild meat with actual market data). Direct observation can be very satisfying. The researcher may feel that because direct observation data are concrete and collected by themselves, they are more valid than data based on the perceptions and opinions of others. However, it is important to remember that these data are only as valid as the researcher's understanding of what is happening. They can be open to misinterpretation. For example, the number of animals brought into the village by a hunter is only a partial count of the number actually killed in the forest, some of which may have rotted in the snares, escaped wounded, or been eaten by the hunter.

Box 3.2 Use of market data—pigs in Sulawesi.

There are two species of endemic wild pig traded in north Sulawesi, the Sulawesi Wild Pig (*Sus celebensis*) and the endangered babirusa (*Babyrousa babyrussa*). Active conservation intervention aimed at halting the illegal trade in babirusas includes checkpoints on the road and awareness-raising with hunters and local villagers. Data were collected from 1993 to 1999 by a local monitor in the main market selling babirusas. She did her weekly household shopping in the market, and while there noted the quantities and prices of meats on sale. Traders were not aware that she was collecting these data. This technique allowed data on pig sales to be collected independently of conservation actions, which made the dataset a useful indicator of the effectiveness of conservation actions.

The data show that although the overall number of wild pigs on sale in the market did not trend over time, there was dramatic variation in babirusa sales. This variation coincided with law enforcement episodes (Figure 3.4).

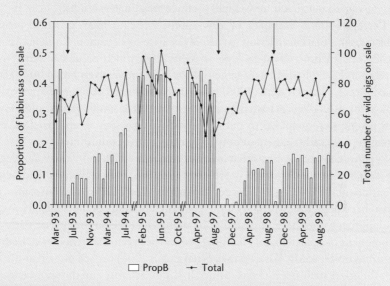

Fig. 3.4. The number of wild pigs and proportion of babirusas on sale in Manado market, 1993–1999. Arrows show occasions on which government inspectors visited the market looking for protected species on sale and/or checkpoints were carried out on the roads. As traders got used to these actions, their reaction became less drastic, with a shorter and shorter period in which babirusas were not sold, as they realised they would not be prosecuted. From 1997 an awareness raising programme began among traders, which seems to have reduced the baseline proportion of babirusas sold.

Source: Milner-Gulland and Clayton (2002).

Collectors may change their behaviour as a result of you being with them, for example, not staying out as long as usual or avoiding illegal activity. Hence, whenever possible, it is important to have supporting information collected in other ways to ensure that what you think you are seeing is what you actually are seeing.

3.2.4.4 Experimental economics

Experimental economics has a similar structure to experimental ecology. Theory leads to hypotheses, which can be tested first in controlled conditions and then in the field (Smith 1994). Much of economics is based on observation—we see a correlation between changes in the quantity of a good on sale and its price and draw conclusions based on our underlying theory of supply and demand. Experimental economics, on the other hand, sets out to test theories by creating a situation in which people can be observed acting out their preferences. It is a specialised branch of economics, and care is needed in experimental design in order to produce valid results. In particular you need a firm grasp of the underlying economic theory that you are setting out to test, or you will produce meaningless results.

There are some examples of this approach being used to study people's behaviour in situations relevant to natural resource conservation. Cardenas (2004) carried out an experiment looking at the way in which natural resource users develop and comply with management rules. In an impressively broad study across 15 countries, Henrich *et al.* (2006) used an experimental economics approach to look at the psychological basis of punishment in a set of simple games, and showed that people were prepared to punish behaviour that they disapproved of, even though it was costly to them personally to carry out the punishment.

Experimental economics often involves games in which individuals are playing for a small amount of money, or which are conducted in artificial, hypothetical situations. This is necessary in order to produce clear quantitative outcomes, but just as in ecology, the artificiality limits the generalisability of the results. However, this approach does have potential to be useful in conservation, particularly as a component of a participatory learning process between researchers and the communities under study. For example, Sirén *et al.* (2006) carried out a real-life lottery for prizes such as a cockerel, tinned fish, poultry wire and shotgun cartridges in an indigenous community in Amazonian Ecuador, aiming to understand the choices that people make between hunting and other livelihood activities. This study was not that informative in terms of providing clear answers to a research question, but it did get the community talking about how income and alternative livelihood activities interact to determine hunting pressure.

3.2.4.5 Using the literature

There is a huge range of literature types available, at all scales from the local to the global. Here are some ideas:

- There are many sources of **official statistics** from governments and international organisations, some of which are freely downloadable from the internet (see the end of the chapter for links). These give exchange rates, inflation rates, import and export figures, human population density, threat categories of species, land cover and much more. It is worth remembering that official data are not always reliable; however concrete they may look in the database, they can, in fact, be very patchy. There is likely to be bias in data quality between countries that could be correlated with the quantity you are interested in measuring. For example, data quality on the number of threatened species in a country is likely to be worst in poorer countries, which are also the most biodiverse. Aggregate figures are not always relevant to local situations. For example, in large and disparate countries like Indonesia, local inflation rates can differ dramatically from the rates in the capital city.

- **Local government offices** are useful sources of relevant data that cannot be remotely accessed. This might include unpublished reports, maps, local economic statistics, data on conservation expenditures. Some of these data are likely to be confidential or difficult to access.

- Obviously there is a need to read all the relevant literature on a study site, including **published accounts** of previous studies, and of studies in other areas that use similar methods. However, in conservation and development there is often a lot of **'grey' literature**—reports that are not published in recognised outlets, such as consultancy reports and conference presentations. While web searches are increasingly a fruitful way to find grey literature, to get a complete picture you need to make personal contact with people who might have copies, or who know the authors. This requires network-building through attending meetings, joining newsgroups and approaching people with knowledge of the area.

- There are also many creative sources of information. For example, explorers' tales can be extremely useful in reconstructing patterns of abundance. Cornwallis Harris (quoted in Martin and Martin 1982) recorded extremely abundant white rhino populations in Cape region of South Africa in the 1830s, but only 70 years later Selous (1908) was writing that over-hunting had led to the imminent extinction of the species. Pandolfi *et al.* (2001) used historical documents to show that reef ecosystems had been fundamentally altered due to over-exploitation long before current concerns about bleaching and disease developed. Caldecott (1988) used official export records of illipe nuts from Sarawak as a proxy for mast fruiting events in the region's Dipterocarp forests. He also used locally obtained records of the products purchased by the Education Authority for children's school meals as an indicator of wildlife abundance. He was able to show that turtles were more often eaten at times of low water levels (when they were easier to catch) and that bearded pigs were more often eaten during mast fruiting events (when they became very abundant through migration and population growth). The famous snowshoe hare cycle was deduced from nearly a century of records of pelts lodged with the Hudson Bay Company (Maclulich 1937).

However useful they can be, these alternative information sources are limited, potentially unreliable, and prone to bias and misinterpretation. The researcher needs to understand the material in order to draw correct inferences. For example, Whitley (1994) discusses the socio-cultural reasons for the rock art he studied being full of representations of hunters killing bighorn sheep, despite the fact that the society was predominately seed-eating.

3.2.5 Techniques for data analysis

As might be expected from its intellectual tradition, there is a wide range of methods for analysing **qualitative data**. These include arts-based analysis in which the researcher collects evidence, thinks about its meaning and then constructs an interpretation based on their experience and understanding. This may be aided by the construction of diagrams illustrating linkages, flows and overlaps between actors and processes.

Much qualitative research is based on detailed **textual analysis** of field notes and interview transcripts (Strauss and Corbin 1990; Coffey *et al.* 1996; Seidel 1998). This highlights patterns in the data, which can then be analysed systematically. There are a number of software packages available that can be used to automate the process (see links at the end of the chapter). The software allows the researcher to search efficiently for particular words or phrases. It can then identify juxtapositions between these words, and hence discover patterns. So for example, a detailed textual analysis of a particular hunter interview may suggest that whenever hunter X is discussing his motivations for hunting he talks about his cultural identity. Then a broader search of all the interview transcripts could be used to confirm that people do indeed seem to link hunting to their cultural identity.

Quantitative data are analysed using **statistical techniques**. However, the analysis should always start with simple data exploration. This includes graphing the data so that a visual impression of relationships is obtained. Then simple univariate analyses can be carried out, such as Chi-squared tests, correlation analyses or t-tests. More complex statistics to analyse the effects of multiple factors and their interactions can then be used, based on the understanding obtained in the exploratory phase of data analysis (Chapter 4). These include general linear models, logistic regression and correspondence analysis. Many statistics textbooks are available, online or in print (see Sections 3.4 and 4.6 for suggestions).

Model-based analyses are also useful tools. In this case, the data are used to parameterise a model, which expresses how the researcher understands the system. For example, a model of hunter behaviour might test the hypothesis that hunters actively conserve their resources by ignoring hunting opportunities in depleted areas (Alvard 1993). The parameterised model is then validated against independent data to show how closely it predicts reality. Rowcliffe *et al.* (2003) used data on prey densities, prey encounter rates with snares and probability of snares catching the prey in a model to predict offtake rates. They could then validate their model using actual offtake rates from the same systems. We explore these kinds of models in more detail in Chapter 5.

3.2.5.1 Cost–benefit analysis

Another kind of model-based analysis assesses the economic costs and benefits of an activity such as hunting. **Cost–benefit analysis** calculates all the different components of costs and benefits that the activity entails, and weighs them against each other to see if the activity produces a net benefit. The analysis is usually done in monetary terms for the sake of convenience, although this is not strictly necessary. Some components of the model are easy to get values for by direct observation or by survey methods (for example, revenues from selling bushmeat, costs of buying cartridges, snare wire or a boat). These tend to be direct costs and benefits, that are already expressed in monetary terms.

Estimates of the monetary value of other costs and benefits must be obtained indirectly. The most important of these is **opportunity costs**. These are the benefits that would have been obtained from the activity that the natural resource user has had to forego because they have limited time. So, for example, many bushmeat hunters are primarily farmers who snare around their fields. In this case, the opportunity costs of hunting are very low, because it is taking very little time away from their other activities. Other people, such as full-time fishers, may be obtaining revenues from fishing at the expense of getting a job in another sector. Opportunity costs are calculated as the wage obtainable in the most lucrative alternative profession. For example, if a hunter is full-time and people of comparable standing and education in the village are farmers or labourers, the profits made by these individuals would be a good estimate of opportunity costs. It can be quite difficult to find a reasonable estimate of opportunity costs in some cases; for example if the hunter is only hunting at night, and the cost is more in terms of reduced productivity in his daytime job. One way to address this is to see if hunting households have a lower agricultural production than non-hunting households, and use this difference as a measure of opportunity cost.

Another cost that must be indirectly obtained is the **cost of being caught** and receiving a fine or a prison sentence. This is made up of two components; the chance of being caught and prosecuted, and the penalty that you are likely to be given if you are prosecuted. Milner-Gulland and Leader-Williams (1992) did a cost–benefit analysis for elephant and rhino poaching that included this cost of illegal hunting, and showed that it was a significant factor in deterring small-scale local poachers from hunting, but it was not significant for the commercial gangs who were most responsible for killing the elephants and rhinos in the area.

Indirect benefits include the cost savings from eating meat from a hunt, when otherwise food would have had to be bought with cash. This can be estimated by calculating how much it would cost a household to buy that amount of meat of comparable quality. Other costs and benefits are more difficult to calculate, because they have no clear monetary value. These include the cultural importance of hunting, or the enjoyment that the person obtains from being out in the wilderness. Conversely, hunting may have a non-monetary cost if people perceive it as dangerous, difficult or lonely. There is a huge literature in **environmental economics** on how to put a value on goods that have no market, and on how to monetise people's perceptions of the value of the environment to themselves and to society (e.g. Bann 1998; Hanley *et al.* 2001).

3.2.5.2 Analysis strategy

As we have seen, there is a huge range of analyses available to the researcher. The kind of analyses that are appropriate depend on the type of data available, and the questions that need to be answered. In the case studies section we make suggestions about how to analyse data on particular topics. In general, it is important to have a feel for the data before ploughing into analyses. Meaningful analyses that lead to understanding are based on testing particular hypotheses and focusing on particular research questions. Hence it is not recommended simply to test a long list of potential explanatory variables for their relationship with a dependent variable. Instead, formulate hypotheses based on your understanding of the system, plot the dependent variable against potential explanatory variables, look for patterns, think about relationships that might exist between variables. And then construct models based on this understanding, which can ideally be validated against independent data. This theme is taken up again in Chapters 4 and 5.

3.3 Case studies

In this section, we have chosen a selection of studies for critical analysis. We highlight the questions that the studies aimed to answer, the data collection and analysis methods that they used, and briefly mention their results (which are not the main focus here). We also give a few ideas for future work that could build on the analyses conducted in the studies. Of course a few case studies will not provide a complete overview of all the possible approaches that could be taken to studying people who use natural resources. Neither do we suggest that you should follow exactly the research protocols used by a particular study. Instead we wish to emphasise the wide range of tools that people use, give some examples of good practice, and start you off on your pre-study literature review.

3.3.1 Individual resource users' behaviour

Much development attention focuses on communities. But in the end, it is individuals who make the decisions about their natural resource use. These decisions range from whether to go hunting, fishing or collecting at all, through what equipment to use, to whether to take a particular animal or plant that they encounter. The first two case studies concentrate on resource user behaviour in a narrow context, looking at how they choose where to go, and whether to use resources legally or illegally.

3.3.1.1 Impact of hunting on wildlife in the Dja reserve, Cameroon

Muchaal and Ngandjui's (1999) paper had two aspects. It recorded hunter behaviour and movement patterns, and it also estimated prey abundance. By combining these two, an estimate of the sustainability of offtake was obtained. Here we concentrate only on the hunter behaviour aspect of the work.

The paper's **research aim** was to study the spatial distribution and intensity of hunting so as to evaluate the effects of hunting on the mammals of the area. This was a component of the development of a management plan for the Dja protected area. Although villagers were hunting inside the reserve, they were working within the village's zone of utilisation, and hence were not acting illegally.

Key elements of the **experimental design**:

- The authors used a Global Positioning System (GPS) to map hunting paths, snare trails and hunting camps of 14 active hunters.
- They divided the area into three zones based on the density of snaring that they found (the lower density zones were further from the village).
- They carried out large numbers of hunter follows (on average 5 days per month per hunter). Hunter follows involve the researcher accompanying hunting or snare-checking trips, and are a useful way of collecting behavioural data. They can be difficult to arrange, because the hunter may be reluctant to participate in case the researcher slows him down.
- They collected data on the animals caught, including age and sex and, importantly, whether the animal had rotted in the snare or not. This measure of wastage is particularly important in assessing sustainability.

The **data analysis** was relatively simple. It involved calculating means by zone and season for snare densities, how often hunters visited their snares, wastage through rotting, and whether animals were sold or eaten at home. The focus of the study was more strongly towards estimating animal abundance than analysis of the hunter data. If the study had been more focused on hunters, then a GLM approach such as used by de Merode *et al.* (2004, see below) would have been a good way to analyse the data.

The **results** of the study include:

- Hunting activity was strongly seasonal.
- Wastage varied by zone. In the zone closest to the village, snares were visited often and at the height of the snaring season only 5.7% of animals were wasted. This compares with 28.5% in the furthest zone.
- Most of the animals captured in the zone near the village were eaten at home, while most of the ones captured further away were sold. This suggests that more commercialised hunters were prepared to travel further.

One avenue for **future investigation** is based the rapid advance of GPS technology. One promising software package is **Cybertracker** (Figure 3.5). This uses a palm pilot to record data which can be automatically georeferenced by a GPS. It can be programmed to allow even non-literate users to enter data, using pictures instead of words. Thus hunters can carry them as they go hunting, and quickly enter data without the need for a researcher to be present. This gives the possibility for much more detailed data collected at a higher spatial resolution and with a larger sample

Fig. 3.5 A research assistant using a cybertracker to record hunter locations, Rio Muni, Equatorial Guinea. Photo © Nöelle Kümpel.

size than is possible if the number of follows is limited by the researcher's presence. One potential problem is getting a good GPS signal in dense forest. Researchers have also had trouble with programming and downloading the data, and with ensuring adequate supervision of the people using them. They are expensive items and vulnerable to theft and damage. However, in the right circumstances, cybertrackers have the potential to revolutionise the collection of hunter data.

A particularly interesting aspect of resource user behaviour is how users perceive and respond to prey depletion. There are a number of options, including increasing their effort (for example, laying more snares) or moving to a new area. The effect that users have on their prey can only be well understood in the context of these decisions. Hence, researchers on resource user behaviour can obtain both qualitative data on how users choose when and where to operate, and quantitative data on the actual encounters that they have when out hunting or gathering. Combining these two data types into a model of resource user decisions will be a major challenge, but one that will improve our understanding of user behaviour.

Muchaal and Ngandjui (1999) note that there is a high level of variation in hunter effort and success. Some of this is due to external factors, such as the season. Some is to do with choices made by hunters, such as the location they hunt in (which determines prey density and hunting costs) and the hunting method they use. There are also likely to be individual differences in skill between hunters. This variation is important and interesting. It also necessitates careful sampling so that all the sources of variation are covered with an adequate sample size. In practice this means that as many hunters as possible need to be followed in different seasons and locations and on as many individual trips as possible.

3.3.1.2 Fuelwood collection in Lake Malawi National Park

The aim of Abbott and Mace's (1999) study was to explore whether law enforcement had any effect on women's fuelwood gathering behaviour. Gathering fuelwood in the National Park was legal so long as women purchased a cheap permit. It was the primary fuel source in the area, and the Park was the only local supply available.

The experimental design involved:

- Following 42 groups of women collecting fuelwood during a 6 month field period, and recording their routes and encounters with game scout patrols. Note that as long as the researcher was not herself collecting wood, she was not committing an offence by being in the Park, and she did not influence the women's behaviour.
- Over the same time-period, scouts recorded the locations of each of their patrols in 1 km^2 grid squares, and recorded any encounters with wood collectors. The small size of the grid squares ensured that any illegal activity was detected in a grid square if it was visited.
- The routes of the women and the scouts were recorded on the same grid squares, so that they could be compared.

Data analysis involved:

- Means were calculated for patrol effort and the encounter rate of patrols with wood collectors. There were two separate estimates for encounter rate, one from the women and one from the patrols, allowing triangulation of the results.
- A qualitative analysis of the spatial location of patrols and of fuelwood trips showed a high correlation between the two. This suggests that patrols were targetting the areas most preferred by the women.
- However, the data also show that patrol effort was so low that the women risked being captured very rarely, even though the patrols were concentrated in the areas where they collected most.
- A cost–benefit analysis was carried out, using a simple substitute cost approach based on the cost of a bundle of fuelwood in the local village (the assumed substitute good for the fuelwood they were collecting). This involved weighing the annual cost of a permit against the annual expected cost of not having one. The expected cost of not having a permit was the chance of being caught multiplied by the fine applied and the value of the fuelwood confiscated. However, most women just received a warning when they were caught, or were asked to buy a permit from the scouts on the spot.

The **results** showed that that 84% of the fuelwood collectors tracked by the researcher were collecting illegally (without a permit), while the patrols recorded 64% of women encountered as having no permit. This discrepancy is suggested to be because illegal collectors tried to avoid encountering patrols, and shows the importance of triangulation. The cost of permits for legal wood collection was US$3.20 per year, compared to an expected cost of US$0.68 in fines and confiscation from harvesting wood illegally. The women were acting rationally in harvesting wood illegally. The detection rate would have had to increase from 12% of a woman's trips to 58% for the balance to change in favour of buying a permit.

The **conclusions** suggested that law enforcement was not the best way to influence women's behaviour towards sustainability, because fuelwood was a necessity. This was already being tacitly realised by the park authorities, in that they had introduced a policy of cautioning the women rather than fining them in order to improve relations with the local villagers.

Future directions: Despite the relatively simple analysis, this study clearly revealed the drivers behind the behaviour of fuelwood collectors in the Lake Malawi National Park. The spatial analysis enabled the behaviour of the users and patrols to be compared and conclusions to be drawn. However, it was a static analysis, in that there was assumed to be no change in the women's or the patrols' behaviour. One extension would be to consider how depletion of the fuelwood resource might change behaviour, such as the locations where it was most cost-effective to gather the wood. This could then be incorporated into an analysis of sustainability.

3.3.2 Natural resource use as a component of livelihoods

3.3.2.1 Background

The **sustainable livelihoods** approach has received a great deal of attention in recent years. One definition is: 'A livelihood comprises the capabilities, assets (including both material and social resources) and activities required for a means of living. A livelihood is sustainable when it can cope with and recover from stresses and shocks and maintain or enhance its capabilities and assets both now and in the future, while not undermining the natural resource base' (DFID 2001). Using natural resources is one of many ways in which people can earn a living or provide food for their household. Hence it fits comfortably into the sustainable livelihoods approach. By looking at natural resource use in a broad livelihoods context we can better understand why certain members of a community use natural resources, how much of their time they devote to it, what they do with their produce, and how policy changes or prey depletion might change these decisions. Livelihoods are dynamic, so that behaviour changes with circumstances. For example, people can be most dependent on wild resources when times are hard. This means that an analysis of 'typical' resource use patterns can be misleading. Instead we have to think about people's behaviour in terms of how they cope with uncertainty and hardship (Maxwell and Frankenberger 1992; Mehta *et al*. 1999).

In **economic terms**, supply of a good and demand for a good are interlinked; producers will only supply a good that people wish to consume. In subsistence situations, production of the good is for the family alone. In commercial situations, the hunter/forager is supplying a good in order to meet the demand from other households. But in all cases there are similar trade-offs to be made; people choose to spend their productive time on hunting or farming, while consumers choose to eat one meat rather than another. These decisions are the basis of economic analysis.

It may appear that the livelihoods approach is philosophically very different to the economic approach. But although livelihoods analysis may involve more conceptual complexities (such as issues of uncertainty and culture), economic analyses also consider trade-offs. Both approaches recognise that natural resource use is one of a number of activities that a person could chose to engage in, the former by placing it within a broad livelihoods context, and the latter by considering opportunity costs. Both also use similar research techniques to collect data, including questionnaire surveys, focus groups and direct observation.

We illustrate the range of approaches using three case studies. The first looks at both the production and consumption decisions of poor people in a village in the Democratic Republic of Congo. The second estimates the effect of prices and income on consumption of wild foods in Bolivia, while the third examines the effects of various factors on consumption and food preferences on Bioko island, Equatorial Guinea. All three are quantitative in their analytical approach, although they vary in the degree to which they use participatory methods in obtaining the data.

3.3.2.2 The value of wild foods to extremely poor rural households in the DRC

de Merode *et al.*'s (2004) **research questions** were:

- Are wild foods valuable in the study community in terms of household consumption and sales? Rationale: this is primarily an agricultural community, so there is a need to establish whether, and to what extent, hunting and gathering wild foods contributes to their livelihoods.
- Are wild foods more valuable at particular times of year? Rationale: It has been previously shown that people use wild foods more at times of shortage of agricultural goods.
- Are wild foods more valuable for the poorest people in the community? Rationale: It has been suggested that a key reason for considering wild foods in any development strategy is that they are especially important for the poorest people in society, and hence any loss of access to them would affect the most vulnerable people.

Key elements of the **experimental design**:

- The final sample size of 128 households was chosen following a detailed pilot study (32 households), and represented 19% of the community. This pilot study was an important and integral part of the research, in that it was used to train research assistants, calculate sample sizes and develop a wealth ranking system, as well as being a pilot for the survey in a more conventional sense.

- Households were chosen for the main study by systematic sampling of every fifth household along footpaths.
- Data were collected with the help of local research assistants, who were trained during the pilot study.
- The study period of 16 months was chosen to ensure that months of agricultural scarcity and plenty were equally represented.

The **data collection methods** used were:

- For assessing wild food consumption: 24-h recall by the person in each household who prepared the food. The respondent was asked to give a detailed account of all food and drinks consumed by the household during the previous 24 h, and where they came from (bought, collected from the wild, gifts). The questions asked were simple, and the recall period was as short as possible. Because the households were revisited numerous times, random variation in consumption patterns on particular days was evened out.
- Participatory assessment of wealth: A group of four key informants visited each household with the first author during the pilot phase. While walking between houses, they informally discussed the wealth characteristics of the house they had just visited. Subsequently each informant individually placed the households into groups of similar wealth. The informants then met and discussed their wealth groupings. Any discrepancies between the informants in where they placed households were discussed, and consensus was reached. Then a list of attributes of each wealth group was drawn up. This list was used in the main study as a template for assigning households to wealth groups.
- Quantitative assessment of wealth: One of the strengths of this study is that they assessed wealth (a key variable for their research questions) in two different ways, and then were able to cross-validate between their research methods. This is particularly important because the participatory wealth groupings were not based on the main sample, but instead were developed using the pilot sample. For the quantitative assessment they collected data for each household on key measures that the participatory assessment highlighted as important wealth indicators (field size, disposable income, non-monetary income, expenditure, capital assets, food reserves). Direct observation was used for field size and capital assets, while income and expenditure came from the recall survey. Capital assets were measured using a formal questionnaire survey on presence of items such as a bicycle, radio, shotgun.

The **data analysis methods** used were:

- A **cluster analysis** was used to check that the quantitative and qualitative wealth rankings were consistent. The authors used the four continuous quantitative wealth measures (incomes, field size, expenditure, assets) as the basis for their cluster analysis. The method of k-means clustering allows you to specify the number of groups that your data should divide into (here, the four participatory wealth classes). The algorithm then divided the data into four groups in the way that that minimised the variation within the groups and maximised the variation between them [see web links at the end of the chapter for more information on the method].

- The correlation between the cluster that a household was placed into, based on quantitative wealth measures, and the participatory wealth group it belonged to could then be assessed. The authors found a correlation coefficient of 0.26, which with 121 datapoints was significant at $P<0.01$. This means that there was still a fair amount of scatter, but that there was clearly a relationship between the two measures of wealth. Notice of course that the qualitative and quantitative wealth measures are not entirely independent, as the variables in the quantitative measure were identified qualitatively. But the statistical analysis gave the authors confidence that both methods were telling them the same thing, and so it was possible to use either as a valid representation of wealth. Note too that the fact that a particular household may be in different wealth classes under different methods is not a problem so long as overall the methods give similar results, and the differences between them are not systematically biased in one direction or another.

- Three continuous **monetary variables** were also calculated for each household: production (the market value equivalent of crops and wild food produced, plus gifts and net profits from sales), consumption (the market value equivalent of all the goods consumed) and sales (market value income from sales of crops and wild foods). These variables were standardised to take into account differences in household size and composition, being expressed in units of US$ per adult male equivalent. This was important in order to ensure consistency within the sample. Note that wealth was measured in monetary units; this allowed more complex statistical analysis to be carried out, at the potential expense of a deeper qualitative understanding.

- Next, the variables, season and wealth rank, were used as factors in a **generalised linear model** of production and consumption. GLMs have the same underlying philosophy as any other regression model. They are available as standard in many statistical software packages. However, it is not straightforward to produce a correctly specified model and then interpret the output. Crawley (2005, 2007) provides excellent guidance on how to carry out GLMs using the free software R.

The approach to data analysis was highly quantitative, despite the participatory way in which the data were collected. Several different approaches were taken to presenting the data. This included using means to give a straightforward expression of the differences between groups. For example, the poorest wealth group produced goods worth only US$0.1 per adult male per day, while the richest group produced US$0.72 per day—still well below the UN definition of extreme poverty as living on $<$$1 per day. They also used flow diagrams to demonstrate the way in

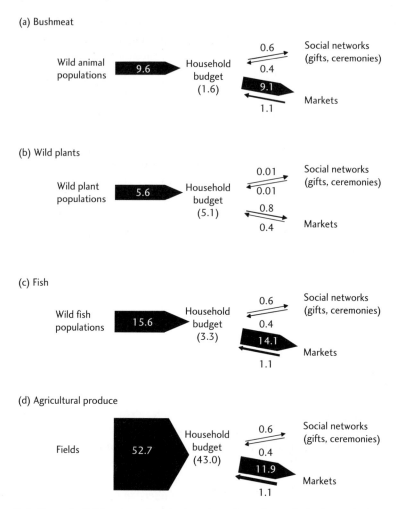

Fig. 3.6 Flow of wild foods and agricultural produce through the household budget. The arrow width and numbers on the arrows represent the volume of flow expressed as a percentage of household production. Household consumption is shown in brackets. The 16% of household production related to industrial products (e.g. oil, salt) is not shown.

which agricultural produce and wild foods came into and left the households (Figure 3.6).

The **results** were related directly to the statistical analyses, and showed that:

- Wild foods were predominately sold, rather than being consumed in the household. Hence they were an important income source for the households.
- Wild foods were more important to households in the lean season, when agricultural crops were scarce.

consumed and sold wild foods to a much lesser
seholds. They seemed to have much lower access to
ket than wealthier households.

rom statistical analysis and towards a more anthropo-
ctors behind the results. The differences in access to
using entitlements theory (Sen 1981). It is only by
nity, and using participatory methods, that this
. Hence this paper is exemplary in showing how the
can be married to produce a fuller understanding
ny one methodological approach on its own. It also
emely clear and focused research questions.

3.3.2.3 Income and price elasticities of bushmeat and fish demand

Wilkie and Godoy (2001) aimed to estimate the elasticities of demand for bushmeat and fish in four indigenous communities in Bolivia. Elasticities of demand
measure the extent to which consumption of a good changes as prices (of the good
itself or of substitute goods) or incomes change. The basic principles of this are covered in Chapter 1. For example, if a new logging company starts operating near a
hunting village, incomes in that village are likely to rise. This might increase their
demand for bushmeat, and so increase hunting rates to supply that demand, with
potential consequences for sustainability.

Elasticities of demand can be estimated in two ways:

- by using a time-series of prices and quantities (for example, from a market);
- by surveying households with different incomes who are paying different
  prices for a good (for example, because they are in different villages). This is
  the method that Wilkie and Godoy use.

There needs to be variation in price, income and quantity consumed in the dataset,
otherwise relationships cannot be distinguished. The time-series method uses
regression analysis to see which factors (e.g. prices, incomes) best explain trends in
quantities purchased over time. The problem with this is that it can be difficult to
assign causation to these correlations (i.e. to be sure that variables are actually influencing each other rather than just co-varying). The cross-sectional household survey method correlates differences in consumption between households with price
and incomes. This approach's problem is that it assumes that current differences
between households are caused by the same processes that would cause changes in
consumption over time. For example, that if a poor household in the sample
became richer, its consumption would change to be like that of the rich households
in the sample. Neither problem is insurmountable, they just require careful data
collection and analysis. Particular attention needs to be paid to missing variables,
which can bias the elasticity estimates. For example, variation in wildlife densities
between villages can affect both price and consumption of bushmeat, distorting
the underlying relationship between the two parameters.

Key elements of the **experimental design** were:

- Four Amerindian groups were visited in Bolivia by two separate researchers, who administered the same household survey. Inter-observer reliability was tested in a pilot study.
- Four hundred and forty-three households were surveyed overall, in 42 villages. The sample proportion averaged 4.3% of the households over the four ethnic groups.
- The male and the female heads of the households were surveyed separately. The analyses were carried out for each of these, but the results were similar, and so the male head of household's results were used in the main analyses.
- Researchers used one-week recall methods, asking the respondent to list all the fish and bushmeat consumed in the week before the interview. This was the dependent variable.
- They also obtained information on explanatory variables, including income from the farm, wage labour and other activities. Wealth was measured by asking whether the household owned each of 12 physical assets. Each of these assets was assigned a monetary value based on its current price in the village being surveyed.
- Other explanatory variables included household size, education level of the household head, village and ethnic group of the respondent, and the price of fish and domestic meat in the village where the household lived.

The **data analysis** included:

- A broad understanding was gained of differences between the four ethnic groups. These included levels of education, degree of isolation from commercial markets, ability to speak Spanish, level of contact with and threat from outsiders. This contextual understanding is important in explaining differences between the groups highlighted in the statistical analyses.
- Six linear regressions were carried out, three for fish and three for bushmeat. These included a regression of consumption on explanatory factors for the sample as a whole and for the top and bottom halves of the income continuum. Dividing the sample into these two income groups helped to elucidate any substantial differences in consumption patterns between richer and poorer people.
- The regressions were carried out using the natural logarithms of all the variables. This meant that the coefficients of the variables were interpretable as elasticities (i.e. the percentage change in consumption given a unit change in price or income). The effects of ethnic group and village were also tested. Only the price and income elasticities were reported.

The **results** showed that:

- About 60% of households ate neither fish nor bushmeat. This led to weaker inferences than if the entire sample had eaten these goods.
- Fish was an inferior good (i.e. the higher the household income, the less fish they ate). This effect was strongest for the lower income group.

- Bushmeat was a necessity good overall, so that the amount eaten increased less than proportionately with income. In the highest income group, bushmeat consumption decreased with income (it was an inferior good). However, none of the elasticities were statistically significant, suggesting that the evidence for an effect of income on consumption was weak.
- Fish consumption was strongly affected by its price in the higher income group, and more weakly related to price in the lower income group. This ability to reduce consumption when price increased suggests that there were plenty of other meats that people could switch to if the price increased. Bushmeat was not sold in the villages so the price could not be estimated.
- Similarly, high chicken and pig prices led to increased fish consumption, suggesting they were substitute goods. Bushmeat consumption did not vary with prices of other meats.

The study's **conclusions** were that changes in income and prices would affect the consumption of fish and bushmeat. The fact that fish was an inferior good is interesting, as it suggests that improving people's incomes would act to reduce pressure on fish stocks. This may also be true for bushmeat, but the results were inconclusive.

Future directions: There have been very few studies of the economics of consumption of wild resources, and this paper represents one of the very few estimates of demand elasticities so far published for wildlife. The policy implications of elasticities are important, so there is much further work to be done on this topic. There is also the interesting issue of how these effects differ between cultures and with differences in market access.

3.3.2.4 Bushmeat consumption and preferences on Bioko Island

A complementary approach to studying actual purchase and consumption behaviour is to ask people about their food preferences. These preferences can then be related to consumption patterns. The role of preferences can be investigated as one factor determining consumption, together with the availability of the product and its price. This is important, because preferences can be powerful drivers of consumption, even if the product is rarely available or expensive, in which case market data cannot reveal these preferences fully. For example, if a particular species is preferred for cultural reasons, then this might lead to continued hunting even if it is very scarce. Alternatively, cultural preferences could be harnessed, with consumers backing conservation as a way to ensure that special foods remain available into the future. This approach has been used in Ghana by the NGO Conservation International, who recruited tribal chiefs to endorse a campaign promoting bushmeat as a part of the country's cultural heritage that is in danger of disappearing (Milner-Gulland *et al.* 2003).

Fa *et al.* (2002) carried out a survey of 196 people on the island of Bioko (Equatorial Guinea), with the aim of quantifying the consumption and preferences of the island's two main ethnic groups. The Bubi are the island's indigenous ethnic group, while the more politically powerful Fang come from the mainland, where there is a wider diversity of bushmeat species available.

Key points in the **experimental design** included:

- The sample was divided into 115 Bubis and 81 Fang. The respondents were located in both urban and rural areas. The sampling strategy was opportunistic, with people chosen according to availability.
- No prior notice was given to the interviewees, although the village chief (or household head, in urban areas) was consulted about the study and asked for permission to carry out interviews. No respondent was given any monetary reward for participating.
- People were asked about their personal preferences and consumption of bushmeat species. They were asked to rank their top three most preferred species and their top three most consumed species from all species together first, and then with the species separated by taxonomic group. Basic information about the respondents (age, sex, location, ethnic group, profession, marital status) were also obtained. There was a broad range covered for each of these variables.
- Preference and consumption scores were ranked as 0 = Not mentioned, 1 = Third place, 2 = Second place, 3 = First place.
- Data on market prices and availability were taken from the published literature.

Data analysis included:

- A preliminary exploration of the data. This involved drawing histograms of the preference and consumption scores, and creating a cross-tabulation to show the relationship between the most preferred and the most consumed species for each ethnic group.
- They used principal components analysis (PCA) to see how the Bubi and Fang differed in their preference and consumption scores. PCA can be useful in reducing datasets with many explanatory variables down to a few composite variables, although interpretation of these composite variables can be difficult.
- They used ordinal logistic regression to find out which variables best explained consumption scores (see Section 3.4 for a weblink). The variables they tested were preference, age, sex and ethnic group.
- They had data on market availability and prices for 11 species, which they used in a multiple linear regression of the effect of a species' availability, price and mean preference score on its mean consumption score.

The main **results** were:

- The Fang named a much wider variety of species than the Bubi, and several of the species they named were not found on the island. The Fang were maintaining their traditional bushmeat preferences, and importing bushmeat from the mainland to satisfy these preferences.
- Generally for the most widely consumed species, consumption was significantly related to preference, as well as to ethnic group. Age and sex were not significant in the regression models.

- For the 11 species with data, consumption was most strongly related to availability. Preference was also significantly related to consumption. Price was not a significant factor in consumption, but the three most consumed species were also the cheapest. These species were also very widely available.

The main **conclusions** were:

- People's relationship with bushmeat was strongly influenced by their ethnic origin. The Fang living on the island still both preferred and consumed mainland species, but Bubis never mentioned these meats. The Bubis preferred species that were commonly available on the island.
- Preferences are related to a complex set of factors, including tastes, traditions, the symbolic meaning of the food, but also to availability.
- One of the main aspects that was missing from the analysis was information on the socio-economic factors affecting consumption. This is important particularly because poverty is likely to co-vary with ethnic group (Bubis tend to be less well off than Fang).

Future directions: There is a lot of potential for further research on consumption patterns and on the structure of trade in exploited species. There are significant differences in prices and consumption patterns between rural and urban dwellers, which probably reflect both preferences and availability (Wilkie *et al.* 2005). Preferences are likely to be dynamic. For example, Ayres *et al.* (1991) showed that consumption of wild meats decreased substantially, and of domestic meats increased when a new road linked an isolated rural community into the wider economy. The degree of fluidity of preferences has an important impact on policies for improving the sustainability of hunting. Analysis of the number of individuals involved in commodity chains, and the profits made along them, is also useful. Bottlenecks in the commodity chain can be useful targets for intervention (for example there are often many resource gatherers and consumers, but only a few traders).

3.3.3 Framework for designing a study of natural resource users' incentives

There is no one-size-fits-all protocol for studying the incentives of natural resource users. The techniques that work best depend on the questions being asked and the characteristics of the population under investigation. Here is one example of a typical framework, for a study of coral collection for construction materials, for sale and for their own use, by a small, poor, rural community. The study would last about 6 months from start to finish.

1. Define the overarching research question.
   - How (un)sustainable is coral collection in the study community, and what can be done to improve sustainability?
2. Define the specific sub-question that will be addressed in this study.
   - What motivates coral collection by households?

- Other sub-questions that could also be addressed in a longer or differently focused study are: How is collection affecting coral populations? (Chapter 2); How might collection levels and practices change in the next few years, with what effect? (Chapters 4 & 5); What institutional changes would help improve the sustainability of local construction practices? (Chapter 6). All four questions would need to be addressed in order to answer the overarching question fully.

3. Define the detailed questions that you will collect data to answer.
   - Do households differ in their use of coral?
     And if so, what factors predict a household's consumption and production of coral?
   - What are the social factors influencing coral use?
     What preferences do people have for using different construction products, wild, cultivated and manufactured?
     How does coral collection compare to other available occupations?
     What is the cultural and social significance of coral as a construction material?
     How do local people view the conservation and management of coral reefs?

4. Prepare a work plan for answering these questions. Make sure all data collected will contribute towards answering your questions (no superfluous data) and that all questions are addressed by the data collected (no missing information).
   - Establish contact with collaborators and local counterparts.
     Prepare fieldwork risk assessment and timetable.
     Obtain permissions from all relevant authorities, including the community head.
     Hire and train local research assistants. These will ideally be from the same area and ethnic group as the study community, but not from the community itself, giving a balance between local understanding and objectivity.
   - Pilot methods with key informants, such as local counterparts.
     Do PRA exercises with groups of informants: community timeline, livelihood matrix, seasonal activity calendar, community mapping, status of and attitudes to coral reef resources.
   - Carry out wealth ranking focus group discussions.
   - Pilot household surveys and revise as necessary.
     Select sample for the household survey, based on the wealth ranking exercise and community map.
     Carry out household survey, collecting information on: household composition, main livelihood activities (both on and off the reef), seasonality of activities, asset ownership, typical use of coral materials, most recent instance of use, typical production of coral materials, most recent instance of production, preferences for construction materials.
     Carry out attitude survey with individuals: their knowledge of the current and past status of the reef and changes in use against the community timeline, their views on conservation and management of the reef.

Collect data on explanatory variables during both surveys (such as age, education, livelihood).

5. If there is an aspect that is of particular interest, and time allows, supplement basic information with further study. For example:

• The economics of coral collection. Length, frequency, location and off-take of collection expeditions; both perceived typical values currently and in the past, and actual observations. Costs of inputs, price and destination of outputs.

• Use picture cards to test recognition of different coral species (including some not found locally as a check on respondents' ability to discriminate) and elicit information on their distribution and characteristics. This could be extended to look at perceived changes in distributions and abundance over time, e.g. the presence of shifting baselines (Saenz-Arroyo *et al.* 2005).

• Market structure. Trace the commodity chain from gatherer to user, including market as well as home use, and obtain details on prices and throughput along the chain. Monitor price and availability of coral and their substitutes locally.

• Investigate the role of coral collection as an activity for lean periods, for example, when fishing is not possible due to the weather, and thus its importance in food security of households.

• Revisit households a number of times, to get short-term recall information on activities since the last visit.

• Collect information on other uses of the reef (such as artisanal fishing) and potential conflicts between different resource use groups. For example, is there a perceived or observable negative interaction between livelihood security of fishers and coral collectors?

6. Analyse data and write up results of the study (shown for data collected under 4 only).

• Validity checks. Is there the expected relationship between wealth rank and asset ownership? Are answers concerning typical production and consumption congruent with direct observations? Is key informant information about seasonality, community history, attitudes congruent with individual household information? Are the data clean, accurately entered and producing the expected patterns? Are there any interesting patterns emerging that deserve investigation?

• Livelihood comparisons. Composite ranking of livelihood activities based on the livelihood matrix exercise. What are the perceived benefits and constraints of coral collection as a livelihood, in comparison with others available?

• Statistical analysis of determinants of coral collection. Coral production and consumption as functions of factors including wealth rank, season, household composition.

• Statistical analysis of preferences for construction materials. Are preferences correlated with availability, price, wealth rank? How does observed use relate to preferences?

- Statistical analysis of correlates of attitudes to coral reef use and conservation. Do older people view coral collection differently to the younger generation, for example?
- Qualitative contextual analyses. How do people view the reef as a resource and component of their livelihoods? How do they view the institutional setting? How do they perceive use to have changed over time? What ideas do they have for improving the sustainability of reef use and their livelihoods?

7. Act on your results.

- Feed back results and recommendations to the community head and to respondents before you leave the study area. Provide a report to them and to local sponsors, in an appropriate format.
- Ensure raw data are made available to local collaborators, with appropriate modifications to ensure confidentiality for respondents.
- Write up and publish analyses in as many forums as possible, so that the results are available to others. Ensure authorship is appropriate, for example, including local collaborators. Ensure you make recommendations for future research and action.
- Think of ways to ensure that your study has a longer-term positive impact. Target policy-makers, support others who wish to continue the work, and hold sessions with local people to discuss ways to address your findings.

3.4 Resources

3.4.1 Websites

Environment and development organisations with downloadable publications:

Institute for Development Studies: www.ids.ac.uk
CIFOR: http://www.cifor.cgiar.org/
International Institute for Environment and Development: http://www.iied.org/
UK Department for International Development: http://www.dfid.gov.uk/
Overseas Development Institute: http://www.odi.org.uk/
Sustainable livelihoods guidance sheets (DFID): http://www.livelihoods.org/
Participatory Planning, Monitoring and Evaluation: http://portals.wdi.wur.nl/ppme/? Tools_%26_Methods
International Development and Research Centre: http://www.idrc.ca/index_en.html
Global Socio-Economic Monitoring Initiative for Coastal management (Soc-Mon): http://international.nos.noaa.gov/socioeconomic/tools.html

Statistics:

StatSoft online statistics textbook: http://www.statsoft.com/textbook/stathome.html
G*Power software: http://www.psycho.uni-duesseldorf.de/aap/projects/gpower/
Simple explanation of statistical power: http://www.jeremymiles.co.uk/misc/power/Cluster Analysis: http://www.statsoft.com/textbook/stcluan.html#general
Ordinal logistic regression: http://www2.chass.ncsu.edu/garson/pa765/logistic.htm Principal components analysis: http://www.statsoft.com/textbook/stfacan.html
R: a very flexible freeware programming language, for use in statistics and modelling: http://www.r-project.org/

Qualitative data analysis:

Computer Assisted Qualitative Data Analysis Software: http://caqdas.soc.surrey.ac.uk/
List of programmes for organising textual data: http://www.lboro.ac.uk/research/mmethods/
   research/software/index-old.html.old
Freeware for textual analysis: http://www.cdc.gov/hiv/software/answr.htm

Codes of practice:

For sociological researchers: http://www.britsoc.co.uk/
For conservationists wishing to publish in Oryx: http://assets.cambridge.org/ORX/orx_ifc.pdf
The Association of Social Anthropologists: http://theasa.org/ethics/ethics_guidelines.htm
American Anthropological Association: http://www.aaanet.org/committees/ethics/ethics.htm

Questionnaire surveys:

Recommended links from de Vaus's textbook www.social-research.org.
A databank of questionnaire surveys: http://qb.soc.surrey.ac.uk

Statistical databases:

The Food and Agriculture Organisation of the United Nations: http://www.fao.org/waicent/
   portal/statistics_en.asp
World Conservation Monitoring Centre: http://www.unep-wcmc.org/
World Bank: http://www.worldbank.org/data/countrydata/countrydata.html
International Financial Statistics Yearbook: http://ifs.apdi.net/imf/about.asp
United Nations Statistics Division: http://unstats.un.org/unsd/
CIA World Factbook: https://www.cia.gov/cia/publications/factbook/
Excellent compendium of data sources: http://www.rba.co.uk/sources/stats.htm
Another good compendium of links: http://www.library.auckland.ac.nz/subjects/stats/offstats/

Other:

Handy currency converter: http://www.oanda.com/convert/classic
Social Sciences Information Gateway: http://www.intute.ac.uk/socialsciences/
Cybertracker website: http://www.cybertracker.co.za/
Royal Geographical Society Expedition Handbook: http://www.rgs.org/OurWork/
   Publications/EAC+publications/Expedition+Handbook/Expedition+Handbook.htm
Society for Conservation Biology Social Science Working Group Catalogue: http://www.con-
   bio.org/workinggroups/sswg/catalog/

3.4.2 Textbooks

Bateman, I., Carson, R.T., Day, B., Hanemann, M., Hanley, N., Hett, T., *et al.* (2002). *Economic Valuation with Stated Preference Techniques: A Manual.* Edward Elgar, Cheltenham. Very useful manual for designing economic valuation studies.

Bernard, H.R. (2002). *Research Methods in Anthropology: Qualitative and Quantitative Methods,* Third Edition. Altamira Press. A useful broad-based textbook.

Crawley, M.J. (2005). *Statistics: an Introduction Using R.* John Wiley and Sons, Chichester. Crawley, M.J. (2007) The R Book. John Wiley and Sons, Chichester. The standard textbook for ecologists wishing to do generalised linear modelling. Many of the same principles apply when you use GLMs in social research. It makes use of R, a popular freeware statistical modelling package.

Hanley, N., Shogren, J., and White, B. (2001). *Introduction to Environmental Economics.* Oxford University Press, UK. A good and widely-used textbook that covers the theory of environmental economics, including valuation of non-market goods.

Kapila, S., and Lyon, F. (1994). *People Oriented Research*. Royal Geographical Society Expedition Advisory Centre. A booklet explaining the basics of social research, aimed at student expeditions. A useful starting point, from www.rgs.org/eac

Patton, M.Q. (1990). *Qualitative Evaluation and Research Methods*, Second Edition. Sage Publications, Newbury Park, CA. An excellent and comprehensive introduction to qualitative research methods.

Schutt, R.K. (2001). *Investigating the Social World: The Process and Practice of Research*, Third Edition. Pine Forge Press, Thousand Oaks, CA. Basic broad text on research methods in social science.

Strauss, A., and Corbin, J. (1998). *Basics of Qualitative Research: Techniques and Procedures for Developing Grounded Theory*, Second Edition. Sage Publications, Thousand Oaks, CA. A classic text book on qualitative research.

de Vaus, D.A. (2002). *Surveys in Social Research*, Fifth Edition. Allen & Unwin, Crow's Nest, Australia. Detailed and readable, covering the full range of topics through theory, design and analysis.

Assessing current sustainability of use

4.1 Scope of the chapter

There are two fundamental approaches to assessing sustainability. First, we can assess past and current sustainability through statistical analysis of our datasets and second, we can use this understanding to predict the effects of future changes. Predictions require the use of mechanistic models, which capture our understanding of the important processes that drive system dynamics. In this chapter, we discuss the way in which we can use data to derive sustainability indicators, and in Chapter 5, we develop models for predicting future sustainability, including the effects of conservation interventions.

There are four main areas for which sustainability indicators can be developed: the biological processes (e.g. species abundance); the interaction between hunters and prey (e.g. offtake rates); the hunter household (e.g. profitability); the social setting (e.g. consumer preferences). These then map onto the facets of sustainability discussed in Chapter 1—biological, social and financial sustainability. We can translate the data we collect into a statistical model that tells us something about sustainability in a number of ways:

- Simple comparisons, for example, with a reference point or between sampling units.
- Regression analyses of trends in a variable linked to sustainability.
- Multivariate analyses of relationships between a set of explanatory variables (which may include time or spatial location) and a dependent variable linked to sustainability.
- Meta-analyses of factors associated with sustainability in a range of studies.

Table 4.1 gives an overview of the kinds of data that could be used to develop sustainability indicators and their pros and cons. We then discuss how to translate these data into indicators using the four types of modelling approach listed above.

4.2 Simple comparisons

We can make **comparisons between sampling units**, for example, between resource abundance in different locations, or between consumption rates in different villages, and ask whether there are statistically significant differences between them. Suitable statistical tests include an analysis of variance, a *t*-test or a

Table 4.1 Potential sustainability indicators.

Note: See Chapter 2 for methods of data collection for the biological data and Chapter 3 for methods to collect the social and harvester data. The column head 'Other data required' refers to requirements for a sustainability assessment. For examples and discussion of each indicator see Sections 4.2 and 4.3.

Indicator	Facet of sustainability	Other data required	Strengths	Drawbacks
Biological				
Population size	Biological	Relation to carrying capacity	Simple measure, can indicate level of depletion.	Can be difficult to obtain. Gives no information on trends.
Population trend	Biological	None—a direct measure	Can use proxy for population size; may be easier to measure than size itself. Direct measure of biological sustainability.	Need a robust and unbiased index, which may be hard to obtain. Need to have adequate power to detect change. Conservation action may be delayed until trend becomes obvious. May indicate trend towards new equilibrium, not unsustainable harvesting.
Population age/sex/size structure	Biological	—	May be more easily obtainable than population size.	Difficult to interpret—best avoided unless strong harvesting selectivity is suspected.
Population productivity	Biological, financial	Population size	Directly relates to the sustainable offtake level at any given population size. Can be estimated allometrically.	Need to assume a functional form for density dependence. Hard to measure in field, allometric relationships may not be reliable.
Species composition of ecosystem	Biological	Trend or comparison with other sites	Measure of broader ecosystem health, may catch effects on non-target species. Can use simple presence–absence data.	Same issues as for population trend. Power to detect change in time to act likely to be low, especially for rarer species.
Hunter–prey interaction				
Number caught	Biological, financial	Population size	Relatively simple to measure—needs to be measured as close to source as possible.	Ambiguous without an estimate of population size. Measures from one point in time can not be used in isolation to estimate sustainability.

Table 4.1 (Con't.)

Indicator	Facet of sustainability	Other data required	Strengths	Drawbacks
Age/sex/size structure in catch	Biological, financial	—	Relatively simple to measure.	Interpretation of results highly problematic, particularly in absence of comparable data on sex/age/size structure in population.
Species composition in catch	Biological, financial	Trend over time or expected/actual composition in ecosystem	Trends in catch composition can indicate extirpation of vulnerable species hence lack of sustainability at ecosystem level.	Interpretation of point estimates requires data on expected or actual species composition in catchment. Trends can be masked if catch comes from areas with different underlying compositions. Power to detect change may be low.
Catch trend	Biological, financial	Population size at one point in time (or trend) and productivity or effort trend	Relatively simple to measure—needs to be measured as close to source as possible. Declining catches may indicate lack of sustainability if all else is equal. Can be normalised for effort if data available. Catch as a function of population size gives harvest mortality, which is a direct measure of sustainability.	Hard to disentangle causality—could be social–economic changes, effort changes or changes in population size, hence the need for independent data on these. May be trending towards new equilibrium rather than unsustainable.
Effort trend	Biological, social, financial	Catch trend	Declining catch per unit effort (CPUE) can indicate depletion. Relatively easy to obtain these data.	Without catch data, cannot be interpreted due to range of potential causes. CPUE does not always have a linear relationship with population size. Effort is multi-faceted, and can be hard to find a suitable metric for.

Table 4.1 (Con't.)

Indicator	Facet of sustainability	Other data required	Strengths	Drawbacks
Catch and/or effort in space	Biological, social, financial	Spatial distribution of effort, catch and/or prey population	Changes in spatial distribution of hunters may indicate a reaction to local depletion (e.g. through increase in distance travelled or shifting between areas).	Causation of changes in hunter effort distribution can be difficult to ascertain—may not be depletion.
Gear types used	Biological, social, financial	Changes in gear type and/or profitability	A component of effort—changes in gear type can cause changes in selectivity between species/sizes and in harvester efficiency and may relate to changes in profitability, hence financial sustainability.	Quite an indirect link to sustainability, needs data on costs and prices (giving profitability) in order to interpret cause and effect.
Hunter household/ village				
Number of hunters	Social, financial, biological	—	Indicates importance of harvesting to household/ community.	Does not inform about biological sustainability without information on catch rates, nor in itself does it inform on dependence on harvesting for livelihoods.
Destination of produce	Social, financial	Trends in catch per unit effort or population size	Proportion sold/gifted/eaten and where sold to can inform about commercialisation and importance to household and society.	Useful for interventions, not so informative about sustainability. Can be supporting evidence for changes in sustainability if trending over time.

Table 4.1 (Con't.)

Indicator	Facet of sustainability	Other data required	Strengths	Drawbacks
Price obtained by harvester	Financial	Harvester costs	Relatively straightforward to obtain.	Need information on costs in order to assess profitability. Care needs to be taken to account for units of measurement (e.g. kg, whole animal) and state of product (fresh, live, processed). If product is mostly eaten, not sold, price may not be useful information.
Harvesting costs	Financial, biological	None—if trend directly related to population size.	Trend in hunting costs can give information on biological sustainability if costs are related to population size. If revenue data available, can infer profitability hence financial sustainability.	Relationship between costs and population size may not be straightforward. Single value not helpful—need trend. Other components of cost need accounting, for example, fixed costs.
Time allocation	Social, financial	Livelihood role of each activity	May help in understanding role of hunting in livelihoods, hence livelihood sustainability. One component of harvester effort (but not the only one).	Time allocation is not the same as importance to livelihoods—may vary in time (e.g. be a fallback in hard times). Too distant from biological sustainability to be of much use except when time is the major harvesting cost.
Compliance with wildlife laws	Social	—	Indicator of institutional sustainability—if laws are not complied with, management is not robust.	Very difficult to get reliable information. Not clearly related to biological sustainability, except as a component of overall offtake.

Table 4.1 (Con't.)

Indicator	Facet of sustainability	Other data required	Strengths	Drawbacks
Social setting				
Consumption levels by households	Social, biological	—	Can relate to both biological sustainability and to social sustainability through contribution of harvesting to livelihoods and food security. Can be triangulated with amounts on sale in market.	Consumption may be a very biased measure of actual harvest mortality if there is significant wastage or unmonitored trade.
Consumer preferences for harvested good and alternatives	Social, financial	Actual consumption	Can be useful in predicting potential changes in hunting levels e.g. through changes in demand for the harvested good.	Preferences are complex to disentangle from price and availability, and may change with circumstances.
Availability of alternative foods	Social, financial	Preferences, prices	Useful for assessing contribution of hunting to livelihoods and options for policy intervention.	Availability is complex to disentangle from price and preferences, and may change with circumstances.
Market price of product and alternatives	Social, financial	—	Price may give indication of market structure, which is of policy relevance. Relatively easy data to obtain.	Price is the outcome of supply and demand dynamics for the hunted good and its alternatives and hence can have multiple causation.

Table 4.1 (Con't.)

Indicator	Facet of sustainability	Other data required	Strengths	Drawbacks
Quantity on sale in market	Social, financial, biological	Population size and productivity.	Relatively easy data to obtain. Gives a minimum estimate of numbers killed. Can be triangulated with household consumption levels. Can be useful in combination with other data.	Quantity on sale is the outcome of supply and demand dynamics for the hunted good and its alternatives, and hence can have multiple causation. It is likely to be a severe underestimate of numbers killed due to losses along the commodity chain. Need to know catchment from which hunted goods came, in order to estimate offtake rates. Multiple harvesting locations with different levels of sustainability may mask trends.
Market species composition	Biological	Trend over time or expected/actual composition in ecosystem	Can indicate extirpation of vulnerable species hence lack of sustainability at ecosystem level, particularly if species not found in the market are of lower productivity than those still found.	Selectivity along the commodity chain for saleable species or changes in gear type may cause changes in market composition. Trends can be masked if catch coming from areas with different underlying compositions. Power to detect change may be low.
Trends in distance travelled to market	Financial, biological	—	Can indicate biological sustainability if all else remains equal. Can be relatively easy to obtain from market traders.	Distance travelled is the outcome of supply and demand dynamics, hence can have multiple causation. Without independent data, population depletion cannot be assumed to be the cause of changes.

Table 4.1 (Con't.)

Indicator	Facet of sustainability	Other data required	Strengths	Drawbacks
Wealth/income distribution of users	Social, financial	Baseline (control population or pre-intervention)	Can be used to monitor social sustainability, e.g. through changes in equitability of wealth. Can also inform about dependence on harvesting among particular social groups.	Can be hard to compare between sites and studies. Not much connection to biological sustainability.
Food security of users	Social	Baseline (control population or pre-intervention)	Can be used to monitor social sustainability. Can also inform about dependence on harvesting in times of need.	Some methodological issues still unresolved.
Health and nutrition of users	Social	Baseline (control population or pre-intervention)	Can be useful to inform policy responses.	Nutritional status is confounded by many factors not just food availability, so analysis must be done with care.
Structure of commodity chain	Social, financial	—	Price changes along the commodity chain can give information on social/financial sustainability. Structure of the chain can indicate robustness of management and be used to predict effect of changes on catch rates.	Not directly related to biological sustainability.
Profitability of harvesting/trading	Financial, biological	—	Can indicate market structure and whether market is at equilibrium.	Involves collection of information on direct costs and revenues and on opportunity costs—quite substantial data requirements.

Table 4.1 (Con't.)

Indicator	Facet of sustainability	Other data required	Strengths	Drawbacks
Available livelihoods	Social, financial	Time allocation, barriers, profitability	The availability of alternative livelihoods and income/consumption from them is useful both for policy intervention and to calculate harvester opportunity costs.	Hard to infer much from this information in isolation.
Levels of conflict in community	Social	Institutional structures, social context	Important information for assessing and intervening in social sustainability.	Not directly related to biological sustainability. May be hard to measure objectively.
Institutional robustness	Social	Social context	Necessary for understanding before intervening.	No clear metric, requires institutions to be tested, which can be difficult to engineer.
Perceived well-being of users	Social	Baseline (control population or pre-intervention)	Can be a useful indicator of social success in conservation interventions.	Complex web of causation, may not be related to harvesting or conservation.
Hunter aspirations	Social	—	Can help in predicting future harvesting pressures and tailoring interventions to needs.	May not be clearly related to current harvesting behaviour.
Attitudes to nature and conservation	Social	—	Can be used to direct appropriate conservation interventions and can be a measure of success.	The link between attitudes and behaviour is not straightforward; behaviour is what leads to changes in sustainability.

chi-squared test. Chi-squared tests are particularly useful for exploration of social data, where variables are often categorical (divided into categories rather than on a continuous scale) and the data are frequencies (such as the number of people saying yes or no to a question). Chapter 3 gives some useful statistical resources, and examples of how comparisons between sampling units have been used in the literature to look at social sustainability, while Chapter 2 discusses statistical methods from a biological perspective.

Another form of comparison is with a **reference point**, defined *a priori* from a theoretical model or from a policy prescription. Examples of social policy prescriptions include that no household must live on less than $1 a day, or every household must have access to fresh water. Biological sustainability is also often assessed against reference points. A commonly used approach when data are scarce is to compare the current catch, effort or population size with a theoretical threshold defining the limit to sustainability (Box 4.1). This idea is the basis of several popular sustainable catch indices (Robinson and Redford 1991; NMFS 1994; Slade *et al.* 1998; Wade 1998; Robinson and Bodmer 1999), although none of these indices is fully rooted in an explicitly defined population model (Milner-Gulland and Akcakaya 2001). While this issue can be overcome, there are a number of more fundamental problems with the use of simple biological indices. These problems don't entirely negate the use of reference points, but it is important to understand the potential pitfalls if you do use them.

First, the interpretation of catch in relation to maximum sustainable yield (MSY) is **ambiguous**. Harvest *less* than MSY may indicate unsustainable offtake from a small, overexploited population (false sustainability), while harvest *greater* than MSY may be the result of sustainable effort during the early stages of a new harvest (false unsustainability). To avoid this ambiguity, you need to know the current population size. Given this information, false sustainability will be defined by a current population below half the carrying capacity. False unsustainability may be suggested, but not unequivocally defined, by a current population close to the carrying capacity.

This also highlights the issue that simple indices assume harvest and population are at **equilibrium**, while the real world is dynamic. One way around this is to avoid basing the comparison on catch alone by using a reference point that expresses catch as a proportion of current population size. This is a safer option because it requires absolute catch to fall if the population declines, and this approach has been adopted by US National Marine Fisheries Service to define limits to allowable by-catch of marine mammals (NMFS 1994; Wade 1998). However, the approach can still be misleading if it is applied at a single point in time. For example, if harvest happens to be unusually light when you make the assessment, you may be unpleasantly surprised later when things return to normal and the population is overexploited. Avoiding this requires continual monitoring and the analysis of trends (Section 4.3).

Reliance on MSY as a reference point provides no buffer against inevitable **uncertainty** in parameter values and offtake estimates, or against **random fluctuations** in

---

Box 4.1 Biological reference points from the logistic model.

The logistic model described in Section 1.3.1.1 allows us to define several possible biological reference points for sustainability. Using the symbols defined in Table 2.1 and Fig. 1.1, the absolute **maximum sustainable yield** (MSY) is given by:

$$MSY = \frac{r_{max}K}{4}$$

Alternatively, MSY can be expressed as a proportion of the current population. This **maximum sustainable proportional yield** is given by:

$$c_{MSY} = \frac{r_{max}}{2}$$

Managing to this reference point is safer because the absolute catch is allowed to fall as the population decreases, but it requires current population size to be known. If harvesting effort is known, but not population size, and it can be assumed that effort and yield are directly proportional (see Section 2.3.5.1 for potential problems with this assumption), then effort can be used as an index of proportional harvest. Given a catchability coefficient, q (the proportion of the population caught per unit effort), the **maximum sustainable effort** is given by:

$$E_{MSY} = \frac{r_{max}}{2q}$$

While catch above MSY causes extinction, proportional harvest above c_{MSY} can in principle be sustained, albeit with lower yield and greater risk to the population. A population is **overexploited** in this way if it is below half carrying capacity, giving the reference point:

$$N_{MSY} = \frac{K}{2}$$

Even a proportional catch can drive a population extinct if it exceeds a certain threshold. These thresholds, the **maximum proportional catch** and **maximum effort** beyond which extinction occurs, are given by:

$$c_{max} = r_{max}$$

$$E_{max} = \frac{r_{max}}{q}$$

Strictly speaking, these equations work only when population production and harvest are both **continuous**—that is, they occur throughout the year at more-or-less constant rates. In practice, the equations are still a reasonable approximation for seasonal systems if r_{max} is low (less than about 0.5), but for more productive seasonal species, the reference points should be based on a **discrete time model**:

$$MSY = K\frac{e^{r_{max}} - 1}{(1 + e^{r_{max}/2})^2}$$

$$c_{MSY} = 1 - \sqrt{e^{-r_{max}}}$$

$$E_{MSY} = \frac{1 - \sqrt{e^{-r_{max}}}}{q}$$

These discrete time equations are conservative in that they give lower maximum catches than the continuous time versions. However, there is no simple population size reference point in this case because the population fluctuates over the year in response to breeding and harvest pulses. N_{MSY} might therefore be substantially more or less than half K, depending on when in the year it is measured.

Simple reference points of this kind are often used to define biological sustainability because they are easy to apply and require relatively little information. The downside of the approach is that it requires strong assumptions about the underlying processes, which are rarely fully justified (see text for discussion of these assumptions). The approach must therefore be used with great caution.

population size. In the management of many commercially exploited species, MSY has historically been interpreted as a goal in itself, usually with the result of disastrous overexploitation because of the unstable nature of the equilibrium at MSY, coupled with natural stochastic fluctuations and flawed information on catch and species biology (Punt and Smith 2001). A common way to deal with this risk is to set the reference point below MSY. For example, Roughgarden and Smith (1996) proposed that an offtake of three-quarters MSY is a robust target for long-term sustainability. However, there is no rigorous basis for defining the proportion of MSY that can safely be caught in any given case. In general, a higher degree of caution is appropriate when the population has a low productivity rate or is naturally highly variable, when catch estimates are imprecise or biased, or when there is little certainty about the basic model parameter values. In the end, though, the choice will primarily be driven by the degree of risk that you are prepared to accept, which cannot be defined objectively.

The reference points in Box 4.1 are based on the **logistic model**. However, the **parameter values** on which the logistic model depends, r_{max} and K, are problematic to estimate, as is q, the catchability coefficient. Although r_{max} has an intuitive meaning, it is difficult to obtain data to estimate it from natural hunted populations, and so it is often estimated by allometry or from simple equations based on survival and productivity rates (see Section 2.4.1). Estimating carrying capacity requires either an allometric approach or data from an unexploited population in the same environmental conditions as the hunted population. This is usually difficult to obtain, because often the reason why an area is unexploited is remoteness or difficult terrain, which is likely to correlate with different habitat types. Allometric relationships are useful, but they are derived from the same flawed data and the variation around the relationship may be both large and biologically meaningful. The catchability coefficient q is usually estimated from data, but as it is not easily observed directly, it is usually necessary to assume an underlying model to derive it, which may be incorrect. All three parameters can be estimated jointly from long-term data on catch and effort or population size data (Section 4.3.3), although the analysis is then focused on trends rather than a one-off comparison.

Finally, **non-linear patterns of density dependence** are thought to be widespread in nature (Sibly *et al.* 2005), and the simple linear density dependence of the logistic model is therefore often likely to be inappropriate. Density dependence is difficult to detect and measure in practice (Brook and Bradshaw 2006; Freckleton *et al.* 2006, Section 2.4.4), and for many species it will be necessary to proceed without a clear understanding of the process. In this case, the logistic model remains useful as a default option; however, it is important to use it in the full knowledge of the potential bias that might result.

In summary, biological reference points are very widely used, despite their problems. The reason for this is that they are simple, they can be calculated when there are only limited data available and they don't need complicated statistics or modelling. This makes them attractive as a 'quick and dirty' way of assessing sustainability. It requires care and understanding to apply them in a way that is still meaningful, rather than being downright misleading.

4.3 Trends over time

The dynamic nature of sustainability means that monitoring trends over time is the most direct way of assessing sustainability. Putting it most simply, if there is a negative trend in a variable that is associated with system sustainability, this is an indicator of concern. Trends are analysed using regression, in which the trend over time in the variable of interest (population size, for example), is related to trends in other variables (for example, number of hunters in the area). A number of issues arise here:

- Is the trend in the variable of interest actually a reflection of **system sustainability**? Trends in prices of wildlife products, for example, usually have multiple causes and are not always directly traceable back to reductions in population size.
- Is any association between the trend in this variable and in other variables actually reflecting **causation**? Might there instead be other factors, not included in the regression, which are impacting on both variables separately or together and so causing a spurious association? For example, perhaps bad weather leads to fishing fleets being unable to leave port, and so a reduction in catch, and at the same time reduces spawning success in that year. This would lead to an association between poor catches and low fish recruitment which has nothing to do with the biological sustainability of fishing, and would only be properly explained if the weather was included in the regression model.
- Is the trend **real**, or is it masked or exacerbated by sampling error or monitoring biases? We discussed this issue in Chapter 2 with respect to ensuring good experimental design and come back to it in Chapter 7 with respect to long-term monitoring.

Sometimes **spatial** variation is used as a proxy for temporal variation. For example, differences in animal abundance between locations that have been hunted for a longer or shorter period of time might be translated into a trend in abundance over time for the area as a whole. The typical example of this in ecology is assuming that succession in time can be investigated by looking at vegetation changes in space (see Begon *et al.* 2005 for a discussion of succession). However, space and time are confounded—for example, people may choose to hunt first in less costly or more productive locations, moving onto lower quality habitats or more distant locations as the earlier ones are exhausted.

It is remarkable how different the language describing regression techniques is between biology and economics. **Econometrics**, the branch of economics dealing with analyses of trends in variables, generally uses different software and different tests of statistical validity to those in common use in biology. If you wish to analyse trends in economic variables, such as prices, over time, it is advisable to think about the particular issues with these types of data, consult an econometrician, and possibly use an econometrics software package such as Stata or Microfit (see Resources section).

4.3.1 Trends in population size or structure

4.3.1.1 Population size

The most direct approach to assessing biological sustainability for a single species is simply to estimate the size of the population on a regular basis and attempt to ensure that it does not consistently decline, and more specifically, that it does not decline below a reference point such N_{MSY} (Box 4.1). However, it is important to remember that a declining population is not necessarily a clear indicator of unsustainable use, nor does a stable population necessarily indicate sustainability. Most obviously, an overexploited population may be stable, but small and therefore at risk. Conversely, when a previously untouched population is first exploited, it declines, but this does not inevitably lead to overexploitation. If the exploitation effort remains constant at an ultimately sustainable level, a new equilibrium will be reached. However, this rebalancing can take a long time—perhaps 10 years or more in species with low intrinsic growth rates. In this case, action to curb offtake may be unnecessary. Given the likely social and economic costs of implementing effort reduction, it would be important to avoid falsely concluding unsustainability in this case. Uncertainty in population estimates can be reduced by looking for trends over a longer time period, but this is likely to be costly and time consuming, with the risk that the population is already overexploited before sufficient data are available to prove it.

Often it is not possible to monitor population size directly. It may only be feasible to monitor **one component of the population**, or to use trends in **relative abundance** as proxies (see Chapter 2). As an example of the former, seals spend a large proportion of their adult lives at sea, and one of the times when it is feasible to count them is as pups (SMRU 2004). However, there are potentially big problems with only monitoring one life stage and assuming that this is a reliable index of total population size (Shea *et al.* 2006; Katzner *et al.* 2007). Taking the example of grey seals in the UK, a change in the number of pups counted could be due to reduced female fecundity or due to increased mortality among juveniles. There is a two-fold difference in population size between these two scenarios; not only that, but because the seal is long-lived it can take years for changes in survival or fecundity to feed through into changes in the population size (SCOS 2006). Any future debate about reopening seal culls to reduce damage to fisheries is likely to hinge on getting an accurate estimate for seal population size.

While it is important to keep these potential problems in mind, **monitoring population size** is probably the most effective method of assessment because it measures directly the variable of interest from a conservation point of view. The more indirect approaches described below provide a means to use a diversity of different information when reliable population abundance information is unavailable. However, in general, the less direct the approach to understanding the current status of a population is, the more assumptions are required, and the more risk arises of being incorrect in the sustainability assessment.

4.3.1.2 Population structure

Sometimes, but not always, exploitation alters the **age structure** of populations. If the harvest is age-selective, the selected ages will become less well represented in the population. If harvest is unselective, the population may or may not become skewed towards juveniles, depending on the nature of density dependence. The extent to which harvest is expected to change age structure, if at all, is thus impossible to generalise. It is particularly hard to define reference points for age structures characteristic of overexploitation. Simple inferences about sustainability from monitoring age structure are therefore rarely possible.

Sex structure can be heavily biased by **sex-selective harvesting**. In the saiga antelope, the ratio of adult males to females went from about 1:5 to 1:100 due to hunting targeting the male horns (Figure 4.1, Milner-Gulland *et al.* 2003). This caused failure to conceive, which was, however, swiftly reversed when the sex ratio rose above about 1:50. There has been concern about the influence of size-selectivity on sex-changing fish; in these species, individuals change sex as they get bigger, and so harvesting large animals may very strongly bias the sex ratio of the population, with potentially serious consequences for sustainability (Platten *et al.* 2002; Molloy *et al.* 2007). However, sex ratios can be naturally highly skewed without affecting population viability, and sex-bias in harvesting has only rarely been shown to have affected population growth. The sex structure of a harvest is a meaningful indicator of sustainability only when the natural sex structure is known, and when the consequences of skewed sex structure for population viability can be assessed.

Trends in size structure are quite widely used as an indicator of sustainability in species such as fish, invertebrates and trees, where both fecundity and profitability are strongly size-related. A large female fish can be many times more fecund than a

Fig. 4.1 Male saigas bear horns which are used in traditional Chinese medicine. Hence they are selectively hunted. Photo © P. Sorokin.

small one, so the actual number of individuals may be less important than their size. Fisheries often target particular size classes (individuals small enough to fit on a plate, for example), and changes in the size structure of the population can affect population dynamics quite profoundly (Jones and Coulson 2006). However, just as for changes in age structure, changes in size structure can be hard to interpret— is a population small-sized due to severe overharvesting of large animals or due to improved recruitment as a population recovers? To answer these questions, more information is needed. The important point here, though, is that the structure of the population is important because it impacts upon its productivity (i.e. the number of new individuals produced per individual present), which determines the sustainable harvest rate. Indeed, some people, particularly in fisheries, suggest that the reproductive potential of a population, modelled using information on both population size and structure, is a better sustainability indicator than abundance alone (Katsukawa *et al.* 2002).

In species such as mammals, where size is less important as a determinant of survival or fecundity, independent of age (at least in females), trends in size structure of the harvest are less informative about sustainability because they are not strongly related to population dynamics. They may, however, indicate long-term genetic changes, which could be significant. For example, Coltman *et al.* (2003) showed reductions in the body and horn size of bighorn sheep due to hunting, which also affected the productivity of the population.

4.3.2 Trends in ecosystem structure

Overexploitation of one or a number of species is likely to have knock-on effects on the rest of the ecosystem. Ecosystem-level effects of fishing are a major field of enquiry in marine systems, where fishing disrupts food supplies, damages habitat and injures or kills individuals of non-target species such as turtles or seals (Tudela 2004). Many governments (including the EU and the USA) are now emphasising the importance of an ecosystems approach to management of fisheries, which integrates fishing into the wider management of the marine ecosystem. One reason why marine protected areas are seen as a useful management tool is that they can address these ecosystem-level effects of fishing through excluding fishers from particular areas (Sumaila *et al.* 2000). In terrestrial conservation, less has been done on the wider effects of hunting, although Redford (1990) coined the term 'the empty forest' referring to bushmeat hunting that removes mammals from otherwise apparently pristine forests. Empty forests may lack the pollination and seed dispersal services provided by animals (e.g. Forget and Jansen 2007), while a study in Boliva (Roldan and Simonetti 2001) showed that heavily hunted forests had far less trampling of tree seedlings than hunted forests, all of which are likely to have knock-on effects on tree recruitment.

It's one thing to demonstrate that hunting has effects on other components of an ecosystem, but quite another to develop indices of sustainability for these kinds of effects, and then use these indices as measures of management effectiveness. However, progress is being made. For example, both the International Council for the Exploration of the Seas (ICES) and the Commission for the Conservation of Antarctic Marine Living Resources (CCAMLR), which are influential international fisheries research bodies, have working groups on the ecosystem effects of fishing activities (see Section 4.6).

4.3.3 Trends in catch per unit effort

Catch data are often much easier to obtain than population size, so it is intuitively appealing to treat **catch alone** as an index of population size and monitor it in order to detect declines. However, inferring sustainability from catch raises an additional problem as well as those involved in using population size directly. The assumption that catch is proportional to population size may be badly mistaken, for a variety of reasons. This can result in a stable catch despite a greatly reduced population. Relying on declining catch to signal overexploitation can in this case be catastrophic, resulting in massive overexploitation before the problem is detected. Conversely, declining catch may simply reflect changing harvester effort driven by social or economic circumstances.

Effort has a very specific meaning in harvesting theory. Rather than the colloquial meaning, it is all the inputs that are put into harvesting. At its most basic, effort is the time spent hunting or foraging, but it also includes the type and efficiency of the weapon used to kill an animal, the mode of transport used to get to the harvesting location, etc. Effort on its own is not a useful indicator of sustainability, because the determinant of harvesting effort is **profitability**. This is made up both of revenues obtained from hunting and costs incurred, which are related both to the size of the harvested population (which we are attempting to estimate) and also to the economic system in which hunting is embedded. Instead, effort is used as an index to standardise catch rates, giving the index **Catch per Unit Effort** (CPUE).

The more abundant a resource is, all else being equal, the easier it is to harvest. Following from this, the greater the CPUE is, the more abundant the population is likely to be. Thus, in principle, CPUE can act as an **index of population abundance**, and could be monitored to detect declines in the same way as abundance itself. This is attractive because catch and the effort put into harvest are relatively easy to measure, compared to biological parameters such as population size, productivity or carrying capacity.

This approach relies on the very strong assumption that CPUE is strictly **proportional to abundance**. There are several reasons why this might not be the case, which are detailed in Section 2.3.5.1. Unfortunately, given catch and effort data alone, there is no way to test the assumption of proportionality, and using these data as a monitoring tool therefore needs to be treated with extreme caution.

As we saw in Figure 1.1, the theory of harvesting predicts that there is a domed relationship between equilibrium catch and effort. Thus, in theory, if we have a series of catch and effort data for widely varying levels of effort, plotting the catch–effort curve may enable us to define a domed response that can be used to define whether a population is overexploited. If current effort and catch are on the right-hand side of the peak, the population is overexploited. This is extremely misleading, however, because it assumes **equilibrium**, whereas data will always be from dynamic systems, with varying levels of effort and populations lagging behind in their responses to changing harvest. If catch and effort from a time series of data from a single location are simply plotted against one another in order to define a maximum, MSY will almost always be overestimated. Using data from several spatially separated populations that have been harvested at contrasting rates could allow this method to work, in principle, but only if each population is close to its equilibrium state, having been harvested at more or less constant rates for a considerable period of time. In practice, such data are very hard to find.

The solution to this problem is to use a **dynamic model** to estimate the crucial parameters. For this, rather than assuming that the population is at equilibrium, one uses the catch and effort time series to model the underlying changes in population size by fitting a dynamic population model to the data (Box 4.2). While this approach is potentially powerful, it is very constrained by the quantity and quality of available data. At least four parameters need to be estimated (intrinsic rate of increase, carrying capacity, catchability and initial population size), and a reasonably long CPUE time series is needed to resolve all of these parameters. Not only that, but the data also needs to contain good **contrast**. This means there should be a lot of variation in the underlying population size over time, and in the amount of effort applied. If you monitor a system where everything is at a fairly steady state, the data will contain no useful information, no matter how long the time series. So the key things that will help you to get meaningful results from a dynamic CPUE model are

- a long series of data;
- more variation in effort;
- monitoring while the system is out of equilibrium (ideally tracking recovery followed by renewed depletion or *vice versa*);
- monitoring a period of overexploitation (although from a conservation perspective this is obviously not desirable).

In addition, the following will also predispose the analysis to producing useful results:

- less natural stochasticity in the population;
- more accurately measured catch and effort;
- prior information on parameter values (for example, an independent estimate of carrying capacity, occasional estimates of actual population size, or an assumption that the initial population was at carrying capacity if harvest was observed from its outset).

Box 4.2 Fitting a dynamic model to catch and effort data.

In the simplest case of catch proportional to effort, we can predict catch at time t on the basis of effort and population size:

$$C_t = qE_t N_t$$

where q is the 'catchability' coefficient, defining the proportion of the population that can be caught per unit effort. We can model population size from one point in time to the next by adding the net growth and subtracting the catch:

$$N_{t+1} = N_t + G_t - C_t$$

where growth, G_t, can be given by any appropriate population growth model; the logistic is frequently used as a default model in the absence of evidence for an alternative structure. In this case, the equation becomes:

$$N_{t+1} = \frac{K}{1 + e^{-r_{max}} \dfrac{K - N_t}{N_t}} - C_t$$

Other possible structures are discussed in Hilborn and Walters (1992). This model can be fitted to a time series of catch and effort data in the same way as this type of data were used over short periods to estimate abundance (Box 2.5), this time estimating the parameters of the population model as well as abundance. Hilborn and Mangel (1997) provide an excellent introduction to the art of model fitting of this kind, while the program CEDA provides an accessible tool for fitting catch–effort models (weblink in Section 4.4).

When the data do not contain sufficient information, dynamic models often give outlandish parameter values with huge standard errors. You should always therefore check that your results make intuitive sense (for example, that the initial population size is about what you would expect) and are reasonably precise.

4.3.4 Species composition of offtake

Where **several species** can be targeted by the same harvest effort, the species with lower intrinsic productivity will be overexploited first, partly because they tend to be larger and more profitable, and also because they are simply more vulnerable. This leads to a progression of local extinctions at any given location (Roberts 1997). Thus, in principle, the species profile of offtake contains information about the state of the harvest system. This is undoubtedly true in broad terms: for example, an offtake consisting entirely of small rodents, where once large ungulates and primates were commonly hunted, almost certainly indicates overexploitation (Rowcliffe *et al.* 2003). This principle has been applied to monitor the state of global fisheries, based on the idea that fishers only turn to species with low trophic level (plankton-feeders) once all the large predatory species are gone. The average trophic level of landings can thus be used as an index of overexploitation. Pauly *et al.* (1998) used this index to show that virtually all marine and freshwater fisheries monitored by the Food and Agriculture Organisation had been overexploited since 1950, especially in the northern hemisphere. This marine trophic index has since been adopted by the Convention on Biodiversity as a means of monitoring progress towards sustainable management of fisheries (Pauly and Watson 2005).

There are some important barriers to the widespread use of this approach. First, the **resolution is low**, requiring fairly dramatic changes in offtake structure before clear patterns can be detected. It is likely that the most vulnerable species will be heavily overexploited or extinct by the time the changes are detected. Second, and exacerbating the first problem, broad spatial coverage in the offtake data (for example, from markets with large catchments) may **obscure local problems**. Widespread overexploitation in some parts of the catchment may be undetected because vulnerable species continue to appear from more recently exploited areas. Finally, if profiles are monitored at a remote end point of trade rather than at the point of harvest, they will reflect the outcome of a chain of transactions, with the potential for considerable **distortion of the original profile** that was harvested. It is hard to generalise about how this might affect conclusions. This approach therefore has potential only as a large-scale monitoring tool for the detection of overexploitation when it has already happened rather than a smaller-scale management tool for the prevention of overexploitation.

4.3.5 Spatial extent of hunting

Changes over time in the distance travelled from base to reach hunting or fishing areas can be related to population depletion. In the extreme case, hunters may empty each area of wildlife and move on, starting with the cheapest, most accessible areas and continuing until there are no unexploited areas left. However, several processes can cause people to travel further to hunt, and teasing out the degree to which overhunting is the cause of observed changes in the distance travelled is very tricky (Box 4.3, Crookes *et al.* 2006).

Box 4.3 Changes in wild pig dealer movements over time.

The accounts over 10 years of a wild pig dealer in Sulawesi showed how he had to travel further and further to obtain supplies of pigs to sell in the market. By 1998 he was travelling for around 24 h as a round-trip to reach the area where his hunters were based, compared to 18 h 10 years earlier (Figure. 4.2).

Although there were indeed fewer and fewer pigs being caught in forests nearer to the market, suggesting depletion, other factors were also at play:

- There was substantial deforestation in the region at the same time, and one of the two wild pig species is confined to primary forests, hence depletion is likely to be due not only to hunting but also to habitat destruction.
- There was a major road improvement during the period, allowing much quicker travel, so that the dealers could go further with minimal additional cost, possibly exacerbating the trend to longer journeys over what would have otherwise been observed.
- There were substantial profits to be made from pig dealing, so that more and more dealers and hunters were entering the market at the time. This competition may have made it worthwhile for the dealer to travel further than others.

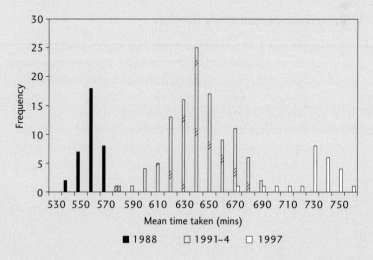

Fig. 4.2 The mean time taken to reach the location of wild pig purchases by one dealer in three time-periods.

- The sales of other commodities, such as dogs and buckets, were increasing, and this may have played a part in the move to longer journeys through increased profits.

Even though intuition may suggest that the trade is unsustainable and wiping the pigs out in a wave of depletion from the market outwards, these other factors need to be considered before a firm conclusion can be drawn. The question then is where will the dealers find it unprofitable to travel further—and will this be before the end of the wild pig habitat?

Source: Clayton *et al.* (1997); Milner-Gulland and Clayton (2002).

It's also important to remember that, just as declines in population size over time may not be unsustainable, but may instead culminate in an equilibrium population size, changes in the spatial extent of harvesting can also **equilibrate** at a sustainable level. It is to be expected that, if harvesters are based in a central location, the area around that location will be depleted, and the areas further away will have higher densities of hunted species. This is not necessarily an indicator of unsustainability. In fact, spatial heterogeneity in hunting pressure may lead to a **source-sink** situation, in which animals are protected in unhunted locations and disperse out into the hunted areas (Novaro *et al.* 2005). We need to recognise that spatial structure in hunting is important for sustainability (Ling and Milner-Gulland in press), but the relationship between lack of sustainability and the spatial extent of hunting is not simple.

4.4 Multivariate explanatory models

Simple regressions of trends in a variable over time are rarely adequate. Many factors may cause an observed trend in CPUE, for example. The prey may simply have become less detectable, perhaps due to the exploitation, leading to biased population estimates. Other aspects of the environment may be changing. This may be associated with the exploitation (for example, logging of tropical forests not only alters the habitat but opens it up to hunters), but may be entirely unrelated (for example, large-scale climatic change can cause populations to fluctuate or decline). Finally, on a shorter timescale, mobile species may simply move out of an area (Boyer *et al.* 2001). The appropriate response to each of these factors would be different.

We might be content simply demonstrating that the observed offtake is sufficient to cause the observed decline. This is not proof of causation, but it invokes the **principle of parsimony**—if one factor is enough to cause an outcome, this suggests that other factors need not be involved. But this is potentially misleading, because interacting factors may cancel each other out and have a minor net effect. It is important to have an understanding of what is actually going on, or conservation interventions aiming to alter the supposed causative factor will not be successful.

So instead, we need to assess the relative contribution of a range of hypothesised factors to changes in our target variable. For example, we may wish to see how differences in both habitat and current hunting effort are linked to population abundance, and how these effects interact with the type of weapon hunters use (Box 4.4). As Box 4.4 shows, data limitations are such that we cannot always disentangle confounding factors without collecting more or different data. But if we start out with a strong set of hypotheses about the processes underlying our observations, we are less likely to be misled by our answers—even if the answer is that we don't know.

This study shows just how difficult it is to disentangle confounding effects on animal abundance. Each species shows different responses to hunting, distance from the village and habitat type—for example, porcupines are associated with rivers and logging roads, but their abundance is not affected by hunting. Many of the primates are not found near the village, even when hunting pressure and habitat type are taken into account, but squirrels are more abundant near the village. Overall, however, animal abundance is strongly negatively related to current hunting pressure.

4.4.1 Confronting models with data

The whole way of thinking about statistical modelling in biology has been revolutionised over the last few years (e.g. Hilborn and Mangel 1997; Burnham and Anderson 2002; Hobbs and Hilborn 2006). This is partly due to the wide availability of statistical computing packages that can deal easily with multivariate analyses, partly due to the rising use of Bayesian statistics, and partly due to the increasing recognition that if science is to influence policy, it needs to present results in a way that quantifies uncertainty in a policy-relevant way. This approach to modelling involves:

- Developing **competing statistical models** which reflect alternative explanations of the processes involved in changes in the variable of interest. The

---

Box 4.4 Disentangling habitat and hunting as determinants of mammal abundance in tropical forests.

Many studies of tropical forest mammals have shown that animal densities increase with distance from human settlement, and this is often attributed to a gradient in hunting pressure (e.g. Peres and Nascimento 2006). However, habitat is also likely to change with distance from a village, for example from agriculture to secondary to primary forest. In a study in Equatorial Guinea, Janna Rist and colleagues tried to disentangle the effects of current and past hunting pressure, habitat type and distance from the village as determinants of the abundance of a range of mammal species (see table below).

Significant minimum adequate model results (with at least one significant abundance predictor), and whether they offer support for hunting (hunt) or habitat (hab) as the most important factor determining abundance, or give no clear signal (–).

Note: 'Survey data' indicates the type of data, and 'Hunting effort' shows the effort measure used in the model. The direction and strength of trend is indicated by positive and negative symbols (one symbol $p < 0.05$, 2 symbols $p < 0.01$, 3 $p < 0.001$), a zero indicates that no significant effect of that predictor was detected.

Species		Survey data	Hunting effort	Abundance predictor					Support for hunting or habitat
				Current hunting	Past hunting	Village distance	Rivers	Logging	
All species		Day sight	All	– –	0	+	–	0	Hunt, Hunt, –
Black colobus	C.satanus	Day sight	Gun	0	0	+++	0	0	Hunt
Moustached monkey	C.cehus	Day sight	Gun	0	+	++	0	0	Hunt, Hunt
Putty-nosed monkey	C.nictitans	Day sight	Gun	– –	0	+++	0	0	Hunt, Hunt
Crowned monkey	C.pogonias	Day sight	Gun	– –	0	0	0	0	Hunt
Gorilla	G.gorilla	Sign	All	– –	0	0	0	0	Hunt
Chimpanzee	P.troglodytes	Sign	All	++	+	+++	0	0	–, –, Hunt
Brush-tailed porcupine	A.africanus	Sign	Trap	0	0	0	++	++	Hab, Hab
Giant pouched rat	C.emini	Sign	Trap	0	0	0	+	0	Hab
Squirrels	Funisciurus spp.	Day sight	Trap	0	0	– –	–	0	–, –
Giant pangolin	S.gigantea	Sign	All	+	0	++	0	–	–, Hunt, –
All duiker species	Cephalopus spp.	Night sight	Trap	0	0	0	+	0	Hab
Blue duiker	C.monticola	Night sight	Trap	0	0	0	+	0	Hab
Red river hog	P.porcus	Sign	All	+++	+++	0	++	0	–, –, Hab
Elephant	L.africana	Sign	All	–	– –	–	++	+++	Hunt, Hunt, –, Hab, Hab
Forest buffalo	S.caffer	Sign	All	0	–	+	0	0	Hunt, Hunt

important thing is that we test not only within a model (for example, by including a range of variables and then simplifying the model by removing non-significant variables in a step-wise fashion), but also consider models with different fundamental structures. For example, we might test a set of models containing different relationships between CPUE and population size (linear and non-linear), and within each model we also have variables for spatial location, habitat type and the gear type used by the hunter.

- Confronting these models with data, and assessing how well each of them performs. This basically means carrying out regression analyses, but the emphasis is on **hypothesis testing**. We first need hypotheses for how the world works (the important factors that affect our dependent variable) and then we can test our predictions using data (Box 4.5).

Box 4.5 Why are Steller's sea lions declining?

The Steller sea lion (*Eumetopias jubatus*, Figure 4.3) has declined dramatically since the late 1970s. There have been many research programmes devoted to

Fig. 4.3 Steller sea lions on Yamski Island, Russia. The western Pacific population of this species declined by 75% between 1976 and 1990. Photo © Tracey Goldstein, the Marine Mammal Centre.

finding out the cause of these declines, but a recent study by Wolf and Mangel (2004) uses all the available spatially disaggregated population data to explore the strength of support for a range of possible hypotheses involving interactions with fisheries, food quality and quantity and predation. This approach is one of an 'ecological detective' (Hilborn and Mangel 1997), letting the data drive the modelling process. They showed strong support for the hypothesis that reductions in the quantity and quality of fish available for the sea lions have a strong effect on fecundity and pup survival, and also that predation by killer whales has a moderate effect on sea lion survival. In this case, exploitation has had an indirect effect on sea lions through fisheries rather than through direct harvest of the sea lions themselves.

Source: Wolf and Mangel (2004).

- Using indices such as **Akaike's Information Criterion** (AIC), which compares between models, giving a measure of which of the models under test is most likely to be true, given the data (Burnham and Anderson 2002). This can then be used as the basis for management, or for understanding where our main uncertainties lie, and so what more information we need. For example, we may find that with the current data, we can not distinguish between the different potential CPUE : abundance relationships, because there are not enough data at the extremes of low population size or catch rate. This is of concern as it means there is a potential risk of population crash as catch rate increases.

We revisit this approach in Sections 5.4.4 and 7.4.2.

4.5 Meta-analyses

As particular approaches to conservation are tried in a range of locations, data become available that can be used to generalise about which approaches work best where, and which factors predispose them to success (Box 4.6). Combining the quantitative results from a number of studies statistically can produce a more robust analysis than looking at each case individually, because the sampling error that afflicts each individual case is reduced. This allows the estimation of the size of the effect of one or more factors on the target variable. This type of analysis is known as a **meta-analysis** (Gurevitch *et al*. 2001). Still rigorous but less quantitative is a **systematic review** (CEBC 2006). Less formal reviews of the available evidence are also useful, and may be more realistic in conservation (e.g. Kellert *et al*. 2000). But in all cases it is important to control for biased sampling of the population of possible studies (for example, project implementers only reporting successes, or reviewers using only examples that support their own world-view) and the effect of different methods and qualities of data collection and analysis on results. In conservation, this is a particular issue, because people are intervening in a messy world, rather than carrying out scientific experiments, so they have less incentive to publish their results, they have varying objectives, and may not all be explicitly measuring the same effects.

Despite the examples in Box 4.6, there is still a long way to go before it is possible to carry out systematic tests of the relative effectiveness of different conservation strategies in achieving prespecified goals. Brooks *et al.* (2006) attempted this and concluded that the quality of reporting precludes full quantitative analysis— 80% of their sample of 124 projects were excluded because the data were inadequate for hypothesis-testing. We come back to the issue of inadequate monitoring of conservation outcomes in Chapter 7.

Box 4.6 Studies of factors predisposing use towards sustainability.

If use is to continue indefinitely, there is a plethora of biological, social, economic and institutional factors that need to be in place. As Table 4.1 showed, it is impossible to choose one variable to measure that can act as a proxy for all aspects of sustainability. A few studies have compared sustainable use projects, trying to tease out the factors contributing to success or failure (e.g. Salafsky *et al.* 1993; Prescott-Allen and Prescott-Allen 1996). Results are often inconclusive because data often cover periods of less than 5 years and many projects are inadequately monitored. The most rigorous study so far is by Salafsky *et al.* (2001). They were looking at use of natural resources in the broad sense (not just hunting but also enterprises such as ecotourism, fruit collection and testing for pharmaceuticals), and at projects that were explicitly community-based. They defined success using an index of the degree to which threats had been reduced (Salafsky and Margoluis 1999), which is indirectly related to sustainability. Their results, based on a statistical analysis of 39 projects, suggested a guide to whether a community-based conservation project based on natural resource use is likely to succeed (see the table below).

Incentive-driven conservation, as examined by Salafsky *et al.* (2001), is not the same as sustainable use; rather it is a conservation philosophy that aims to involve local people in conservation by making it economically worthwhile for them. Sustainable use, on the other hand, can come from any kind of intervention, not just setting up a project designed to make nature pay, and may or may not be community-based (Hutton and Leader-Williams 2003). However, the table gives a flavour of the broad range of issues that need to be considered when deciding whether use is likely to be sustainable. It is interesting to note that biological factors come in only in an indirect way (biological productivity is one determinant of profitability, for example). Instead, factors such as the robustness of social structure and potential for rapid tangible benefits are highlighted.

A guide to whether it is worthwhile proceeding with a community-based conservation through use project.

Note: 'Bad prospects' on one factor does not necessarily rule out success, but means that there should be some hard thinking before starting on the project.

Factor	Good prospects	Bad prospects
The enterprise		
Profitability	More than covers costs	Does not cover costs
Market demand	Moderate	Too high or too low
Infrastructure	Good	Poor
Local skills	High	Limited
Complexity	Low	High
Linkage to conservation	High	Low (or not perceived)
Benefits		
Cash benefits	Moderate	Low or too high
Non-cash benefits	High	Limited
Time benefits received	Immediate	Long or uncertain wait
Distribution	Targeted to those conserving	Too broad or to wrong people
Stakeholders		
Group of stakeholders	Established	Absent or weak
Leadership	Balanced and respected	Absent or too strong
Access to resource	Full	None or ill-defined
Enforceability of rules	Strong	Weak or none
Homogeneity of group	Complete	Limited
Conflict	Absent	Present
Threat source	External	Predominately internal
Other		
Chaotic situation	Unlikely or can be adapted to	Endemic
Project alliance	Experienced and established	Otherwise

Source: Adapted and simplified from Salafsky *et al.* (2001).

Another example of this kind of analysis was carried out by Halls *et al.* (2002) for the Department for International Development, UK. DFID had supported many small-scale fisheries co-management schemes in developing countries, and wanted a quantitative assessment of success of these schemes. Data were available for 258 variables that might affect sustainability from 119 fisheries in 13 countries. Halls *et al.* (2002) used a range of statistical techniques to approach the issue, starting with principal components analysis and cluster analysis to make the number of explanatory variables more manageable (Chapter 3). Then they built general linear models (Crawley 2005) to describe the relationship between a range of proposed explanatory variables and three dependent variables, catch per unit area and effort, and household income. The strongest relationships they found were between catch per unit area/effort and density of fishers, which varied geographically. They then went on to develop a Bayesian network model (see Box 5.4) to investigate the factors affecting equity of income distribution and compliance with rules (both likely to be important components of social sustainability), as well as CPUE (see the table below).

Factors found to be important in a preliminary analysis of sustainability of artisanal fisheries.

Note: Equity is the evenness of distribution of fishing yields (high or low). CPUE change is whether the CPUE is declining or not, and compliance is a subjective measure of the level of compliance with the management rules (high/low). Management type is either government or co-managed/ community-managed. The number is the rank order of importance in determining a positive outcome for a particular sustainability indicator (e.g. conflict resolution has the highest effect on equity). Relationships are positive unless otherwise noted (e.g. more gears = higher equity, more effective control and surveillance = higher probability of CPUE being stable or increasing). Source: Halls *et al.* (2002).

Factor	Equity	CPUE change	Compliance
Effective conflict resolution mechanism	1		
More gears in the fishery	2		
High fisher representation in rule-making	3	2	2
Not government-managed	4	4	4
Democratically elected decision-making body	5	5	5
Effective control and surveillance		1	1
High fisher density		3	
Clear access rights			3

4.6 Resources

4.6.1 Websites

Econometrics packages:

Microfit—for time-series analyses: http://www.econ.cam.ac.uk/microfit/
Stata—the statistics package most economists use: http://www.stata.com/

Databases for meta-analyses:

http://www.conservationevidence.com/
Centre for Evidence-based Conservation: http://www.cebc.bham.ac.uk/ introSR.htm

Fisheries websites:

CEDA software Catch Effort Data Analysis: http://www.fmsp.org.uk/ Software.htm

ICES working group on ecosystem effects of fishing activities: http://www.ices.dk/iceswork/ wgdetailace.asp?wg=WGECO

CCAMLR Ecosystem Monitoring Programme: http://www.ccamlr.org/pu/E/sc/cemp/intro.htm

Good on indicators:

Cochrane, K., ed. (2002) *A fishery manager's guidebook: Management measures and their application.* Food and Agriculture Organisation Technical Paper number 424. FAO, Rome. ftp://ftp.fao.org/docrep/fao/004/y3427e/y3427e00.pdf

4.6.2 Textbooks

Begg, D., Fischer, S., and Dornbusch, R. (2005). *Economics*, Eighth Edition. McGraw-Hill, London. Classic textbook for undergraduate economists.

Burnham, K.P., and Anderson, D.R. (2002). *Model Selection and Multi-Model Inference: A Practical Information-Theoretic Approach*, Springer-Verlag, New York, NY. The Bible for the new approach to model selection.

Gujarati, D. (2003). *Basic Econometrics*, Fourth Edition. McGraw-Hill, Sydney. Very widely used text for undergraduate economists.

Hilborn, R. and Mangel, M. (1997). *The Ecological Detective: Confronting Models with Data.* Monographs in Population Biology 28. Princeton University Press, Princeton, NJ. A must-have for all who wish to carry out quantitative analyses based on empirical data. A landmark work.

Maddala, G.S. (2001). *Introduction to Econometrics*, Third Edition. Wiley, New York, NY. Very clear and accessible econometrics textbook.

Developing predictive models

5.1 Scope of the chapter

Analyses of the current situation are useful, but they are not so good at helping us to predict what the effect of an intervention might be, and particularly what the magnitude of that effect might be. If we wish to manage cost-effectively, then it is important to be able to weigh up the outcomes of different options. This requires a mechanistic understanding of the system, not just correlations between variables, which is what the statistical models discussed in Chapter 4 give us. In this chapter we give some tips on building and using a mechanistic model of your system. This usually involves expressing your understanding of the system in the form of mathematical equations, although conceptual models can also be useful, either as a first step or as an end in themselves. A good model is a major contribution to sustainability analyses because it:

- Forces you to be **explicit** about your assumptions concerning how the system works, because you have to write them down in equation format.
- Allows you to investigate the **logical consequences** of the assumptions you have made and parameter values you have chosen for systems that are too complex for intuitive assessment.
- Allows you to investigate the consequences of making **different assumptions** about how the system works and of varying parameter values.
- Gives **transparency** to decision-making processes because every step is explicit.
- Allows you to give a quantitative assessment of **uncertainty**, and how it affects predictions.
- Enables you to generate **hypotheses** about the effects of management interventions, which can then be tested, both within the model, and with empirical data. This forms the basis for adaptive management (Chapter 7).
- In situations when real-life experiments are not possible (a common situation for species of conservation concern) a simulation model allows **experimentation** in a 'virtual world'—see Chapter 7.

To be successful in modelling you don't necessarily need to have high-level mathematical skills. You do need to have patience, logic and a methodical approach, and most importantly not to be satisfied with something that is nearly good enough. You need to convince yourself at every step that you know exactly what is going on, using graphs and other model outputs to check this. You also need time—learning the craft of modelling can't be done as an add-on to a field-based project; it needs to be seen as a research component in its own right. Producing a publication-standard modelling study takes 6–12 months of hard work.

5.2 Types of model

It can be very daunting to embark on modelling from cold, especially if you don't have any other people around you who have done it before. However, the process can be broken down into small steps, and even only travelling a few steps down the path to a fully functioning model is still very beneficial, because it forces you to think clearly and logically about the problem. One of the problems is the jargon involved in modelling. Table 5.1 gives a short guide to terms used to describe the main types of model.

Table 5.1 Common types of model.

Model type	Explanation
Conceptual vs. mathematical	A conceptual model shows qualitatively your understanding of the system, often in a flow diagram. A mathematical model quantifies the relationships between the components of the flow diagram using equations.
Analytical vs. simulation	An analytical model is a set of mathematical equations which are solved to get general relationships between parameters. In a simulation model, specific values are given to the parameters in the equations and the model is run a number of times to get the output for these values.
Equilibrium vs. dynamic	An equilibrium model is a representation of the state that a system will end up in. A dynamic model represents system changes over time, not just the final resting point.
Lumped vs. structured	A lumped model represents the object of interest (e.g. a population) as one number, assuming all individuals are the same. A structured model divides the population up into compartments—sexes, ages, size classes, spatial locations, individuals, or even genotypes.
Continuous vs. discrete time	Models in continuous time assume that all processes happen continuously, while those in discrete time break time down into time-periods, such as years or seasons.
Deterministic vs. stochastic	In deterministic models parameter values are constant, but in stochastic models they are subject to random variation.
Bio-economic vs. others	A bio-economic model involves both the human and the biological components of the system—this is required for a full understanding of the sustainability of the system.

It is important to recognise that a model is a highly simplified representation of reality. Making a model that is so detailed that it captures reality near-perfectly is a pointless exercise. On the other hand, there is a lot of justifiable criticism and suspicion about models in the conservation world, which comes from modellers abstracting so far from reality that their results are not practically useful. So there is a trade-off between **simplicity and realism**. Too simple, and the model doesn't capture the key elements of the system and so it can't predict effectively. Too complex, and the model becomes a black box, in which you have no understanding of the reasons behind results—which negates the purpose of the model in the first place, and is very dangerous when you come to make predictions.

The first step is to create a conceptual model of the system. This in itself will be useful in clarifying your assumptions about how the system works. But it can't be used to quantify the effects of changes in the system or in parameter values, or of uncertainty, on the outcome. This means it is very limited as a predictive tool—it is more like an expression of the understanding that the statistical models in Chapter 4 would give you. Hence it is preferable to go on to produce a **mathematical model**.

Analytical models are of limited use in most of the situations that we are covering in this book. They need to be very simple in order that they can be solved, so bio-economic systems are usually too complex for meaningful analytical models to be built. For example, the simple models used to obtain reference points in Box 4.1 are analytical models. These models also require mathematical skills to solve them. **Simulation** models have the weakness of only telling you about the solution to the model under the particular set of parameter values that you have used. We will focus on simulation models in this chapter—but bearing in mind that the ability of analytical models to produce a mathematical solution to a problem is a big strength.

Equilibrium models are static—they don't take time into account. They are useful inasmuch as it is important to know whether exploitation is tending towards a sustainable equilibrium or to population extirpation. However, bio-economic systems are dynamic, meaning that they need to be modelled over time. There are inherent time-lags through, for example, the effects of hunting on population dynamics and the effect of price and cost changes on hunter behaviour. There are also constant shocks to the system moving it away from equilibrium, such as changes in the weather affecting animal populations, as well as deterministic trends in important variables that are constantly changing the equilibrium itself (such as habitat loss, human population growth or technology improvements). Equilibrium models can be solved analytically, but we will focus on **dynamic** simulation models. These are more flexible, realistic and also easier to build and run because they don't require advanced mathematical abilities—just the ability to correctly specify an equation.

The logistic model in Chapter 1 was a **lumped** model, in which the population size was given the symbol N, and all individuals were implicitly assumed to be the same. However, populations are in reality made up of different components, whether they are populations of animals, plants, hunters or consumers. The extent to which we need to characterise this variation depends on the problem we are addressing. In general, the simpler the model, the better, so **structure** should only be introduced if it is fundamental to the dynamics of the system (Section 5.4.1). For example, it is usually only necessary to include age and sex structure in biological models of population dynamics if the harvest is strongly biased towards a particular age or sex. Individual-based models are computationally very intensive and are not often needed in this field. **Spatially explicit models**, on the other hand, are useful because hunting pressure is often spatially heterogeneous, and the results of spatially explicit models can vary quite substantially from non-spatial models.

Continuous time models are usually used when an analytical solution is required, because they can be solved using calculus. However, they are more difficult to conceptualise and to implement in a simulation model. For this reason, we tend to use **discrete** time models—these also often make more sense in bio-economic systems when time is naturally divided. This is usually by season (open vs. closed hunting seasons, agricultural seasons, fruiting or birth periods) or by year. **Deterministic** models are easier to parameterise and interpret, and in many cases they give an adequate representation of the system. Including random variation (such as weather-related variation in survival) is more realistic, and can make a big difference if the population is small or the variation is large (for example, catastrophic winter mortality can cause extinction in small populations). Another strength of **stochastic** models is that they can be used for risk assessment, while deterministic models only give the most likely outcome based on a fixed set of parameter values.

Finally, we can create **bio-economic** models, that attempt to model the sustainability of the system as a whole, or we can model components of the system separately. For example, we might want to find out what the sustainable level of offtake is that a given population size can provide, using a population dynamics model. Or we might wish to use a cost-benefit model to find out how likely it is that a hunter will decide to break the rules governing hunting, based on his perceptions of the likely punishment and the profitability of poaching. The trade-off between complexity and predictive power comes in here again, and in general it is important first to determine the **key questions**, and then build a model appropriate to answering these questions.

5.2.1 Off-the-shelf packages

There are a number of packages available that enable you to enter parameter values into a pre-coded population model, and so save yourself the time and effort of building your own model (see Resources section for websites). The strengths of off-the-shelf packages are that they are quick and reliable and give a higher level of analysis than is possible for a first-time modeller. If you are not confident in the principles of population dynamics, then when building your own model you may well make fundamental mistakes in parameter estimation and equation specification that you will not spot. Some packages, like RAMAS, allow you to build quite sophisticated spatial models, combining GIS data with population models, which is something that is very difficult to do well without substantial modelling expertise.

Often the graphics and sensitivity analyses that the models produce automatically are high quality and more extensive than you might be able to produce yourself.

The weaknesses include that the modelling package is only able to answer a limited range of questions, which may not be the ones that you are interested in. For example, all the high profile conservation packages focus on the biological side of things. There are some quite specific forestry and fisheries packages that focus on the manager's perspective, although one, ParFish, is a tool for setting up and managing participatory fisheries, so has a broader perspective. There is a disconnect between toolkits for modelling population dynamics and decision support tools, which don't necessarily include a proper mechanistic model of the system. This lacuna is where many of the questions we are addressing in this book lie. Even if you want to produce a standard population model, the specifics of your system may not fit well with the options that the software provides, for example, the functional forms for density dependence. The model may require information that you don't have (for example, on frequency of catastrophes or genetic diversity), and guessing at values may produce misleading outputs.

In the next Section (5.3) we outline how to build your own model. A self-built model will produce a deeper understanding of your particular system, and modelling is a transferable skill that it is worth investing time in obtaining. Whether or not you use a package, the steps outlined in Section 5.3 are still necessary. It's tempting to use the package as a prop, carrying out the analyses that it suggests without first thinking through your own hypotheses. The package is a short-cut for the actual model-building component of the analysis, but you still need a conceptual model first, data, and an understanding of the underlying processes and the uncertainty involved. You then need to carry out sensitivity analyses and model validation. You also need to read and fully understand the users' manual so that you can enter data appropriately and choose the right model structure.

5.3 Building your own model

In this section, we will go through the stages of model-building, using two simple contrasting models as examples. One is an age-structured model of red deer (*Cervus elaphus*) population dynamics, based on the model presented in Milner-Gulland *et al.* (2004), the other is a very simple bio-economic model of village hunting.

5.3.1 Conceptual model

Even if you go no further, it is worth building a conceptual model of your system as a way of expressing what you know about it and clarifying your logic in thinking about potential interventions. One common form of conceptual model is a flow diagram (Figures 5.1 and 5.2). A few things to bear in mind with these are:

- What do the boxes and arrows actually represent? It is important to separate **processes** and **quantities**, and often the arrows represent processes and the boxes quantities. Do the boxes and arrows represent the same thing in all parts of the diagram? If not, then this needs careful thought to check that the model is logically constructed. It might be worth considering a different shape for some of the boxes—e.g. rectangles for quantities, diamonds for decision points.

- The model is sometimes simply a representation of how some quantity moves around a system (like energy in food web diagrams, or individuals in population models—see conceptual model example 1 below). Other times it represents **causal links**—events A, B and C lead to a decision X, which has effect Y (see conceptual model example 2 below). It is important not to mix up the two kinds of diagram.
- It's useful to run through the diagram in both directions, forwards and backwards, and check that the logic holds whichever way you go.
- Ask yourself whether each step of the model translates into an equation of the form $y = f(x)$, i.e. the outcome box is a function of the input box(es). If yes, then the conceptual model is usable as a basis for a mathematical model.

The conceptual model is a summary of your current understanding of the system, informed either by the literature, or by the data collection and analysis we've discussed in Chapters 2–4. At this conceptualising stage another important step is to think about the uncertainty in our understanding of the system, and whether there are any feasible **alternative model structures**. The outcome of a model is fundamentally determined by the structure of its causal links, so if only one conceptual model is used, this has already to a large extent determined what your answers are. This comes back to the issue of confronting models with data discussed in Section 4.4.1. In population models, for example, there are usually several potentially valid forms for density dependence and dispersal functions. Similarly, in economic models, we make assumptions about people's attitudes to risk and the relationship between money and utility (see Section 5.4.2), which may or may not be valid.

We could also consider using two different types of model and comparing their output, for example, comparing a simple lumped model with an age-structured model. This will tell us whether the added complexity of the structured model leads to a better fit to the data or a worse one (see Section 5.4). Another fundamental question is whether we are including the right variables in the model? Is there a **missing variable**? For example, habitat type may determine animal abundance as well as hunting pressure (Box 4.4).

The most important component of developing a conceptual model is to get a really good understanding of the system through reading and/or fieldwork. Reading as many papers as possible on models developed for similar systems to yours will give you ideas about the type of model which is useful for addressing your research questions.

5.3.1.1 Conceptual model example 1—Red deer

Figure 5.1 shows the flow of individuals through different age classes. We have modelled three age classes using the symbol N— year 1 (juveniles), year 2 (non-breeding sub-adults) and adults (all other ages). The symbol S represents

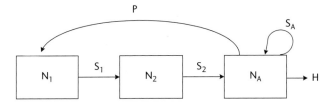

Fig. 5.1 A flow diagram summarising a simple biological system: age structured population growth of a hunted deer population. Boxes represent quantities, while arrows represent rates of flow.

survival from one year to the next—note that adult survival feeds back into the adult box. Adults have a fecundity P, which produces the next year's individuals in the Year 1 category. Adults are also subject to a hunting offtake, H. This model is dynamic, i.e. it tracks a population over time, but here we have shown time only implicitly. We have also ignored dispersal, assuming that we are dealing with a closed population, with no movement of animals in or out other than through births and deaths.

Note that we have put symbols in each of the boxes representing the quantity (in this case the number of animals in an age-class), and next to the arrows representing rates (survival, hunting and birth rates). This will make it easier to turn the conceptual model into mathematical format.

5.3.1.2 Conceptual model example 2—village hunting

This is a model of the dynamics of hunting for a village of a fixed size exploiting a closed population of a single species (Figure 5.2). Individual villagers choose whether to go hunting or not depending on whether the price offered by the

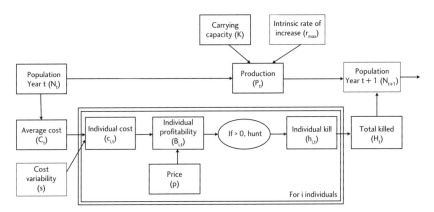

Fig. 5.2. A flow diagram summarising a simple bioeconomic system: the decisions and impacts of village hunters. Rectangles represent variables and constants, the oval is a decision point, and arrows represent causal links.

trader is higher or lower than their cost of hunting (which includes opportunity cost, direct costs of equipment and the cost of the effort expended on the hunting trip). Each hunter kills one animal per hunting trip. The hunted population has logistic growth (Chapter 1).

In this model, we have two kinds of box—one that represents parameters and variables of the model (rectangle) and one that represents decisions (oval). The parameters which are constant throughout have one symbol (K, r_{max}, s, p), while the variables, which vary from year to year, have a subscript representing time (N_t, C_t, P_t, H_t). Finally, the individual cost, $c_{i,t}$, kill, $h_{i,t}$, and profitability, $B_{i,t}$, have two subscripts, meaning that they vary both between individuals, i, and between years, t. The double lines around the boxes for individual variables indicate that the process has to be repeated for each individual hunter and then the number killed summed to get the total. This model just shows the process for one year, but the arrow onwards shows where the next year starts. For clarity of exposition we have the biological component of the model in the top half of the diagram, and the economic component in the bottom half.

5.3.2 Writing the model in equation form

The next step is to turn the conceptual model into equations which can be parameterised from data and used in a model. This should be an easier task because we have already introduced notation into the conceptual model. It's very helpful to use standard notation if you can (e.g. N or x are most commonly used for population size, t is commonly used for time, and i and j for age, stage or sex classes). Use single symbols if at all possible, with subscripts for different components (i.e. N_t is population size at time t), rather than names or two-letter symbols, because these can be confusing. This may all sound petty, but it's easy to make mistakes with sloppy notation, as well as making it hard for others to follow what you're doing.

5.3.2.1 Deer mathematical model

For demographic models, the classic approach is to use a **Leslie matrix**. Getz and Haight (1989) provide a particularly accessible introduction to matrix modelling for harvested species. The Leslie Matrix summarises the transitions in the flow diagram and uses matrix algebra to get the next year's population size as a function of this year's. The equation is:

$$\mathbf{N}_{t+1} = \mathbf{A}\mathbf{N}_t$$

where $\mathbf{N}_t$ is a vector of the number of individuals in each class, and $\mathbf{A}$ is the transition matrix:

$$\begin{bmatrix} 0 & 0 & P \\ S_1 & 0 & 0 \\ 0 & S_2 & S_A(1-H) \end{bmatrix}$$

The matrix is 3×3 because there are three age classes in the model. The top row is the contribution to fecundity—here only the adult class contributes to fecundity, so its cell is the only non-zero one. The sub-diagonal shows the proportion of an age-class moving to the next age-class through survival (S_1 and S_2), the diagonal shows the proportion staying in the same age-class $S_A(1 - H)$.

Leslie Matrices have useful properties. It is possible to calculate values associated with the matrix using a programme like Matlab or R; the dominant eigenvalue, which is the population growth rate λ, and the left eigenvector, which is the stable age distribution when the population is at equilibrium.

The next issue is whether each of the cells of the matrix is just a number or a function itself. Density dependence is an important part of population dynamics, which can only be ignored for very small populations which are growing at their maximal rate, not limited by resources (see Sections 1.3.1.1 and 2.4.4). In order to incorporate density dependence, you need to decide which of the **vital rates**, i.e. the cells of the matrix, are involved—is it juvenile survival S_1 that is affected when resources become limiting, or do adults have fewer offspring, P? In Milner-Gulland *et al.* (2004), the form of density dependence was determined empirically by fitting curves to data, and the shape for fecundity, for example, was:

$$P_t = \frac{1}{1 + e^{(-a + b \cdot N_t)}}$$

i.e. the fecundity in a given year is a non-linear function of the total population size in that year, where a and b are constants. Once density dependence is involved, the model becomes more realistic but the eigenvalues and eigenvectors are no longer easily calculable. This means that instead of finding an analytical solution using the Leslie Matrix, a simulation approach is often used instead. The model is coded and run as a series of vectors, cells of which are consecutively multiplied by the appropriate vital rate (see below). The Leslie Matrix is still a useful way to present the model structure, however.

5.3.2.2 Hunting mathematical model

The hunting model has a number of components. First the biology:

$$N_{t+1} = N_t + P_t - H_t$$

$$P_t = r_{\max} N_t \left(1 - \frac{N_t}{K}\right)$$

This says that the population size next year is the population size last year plus the productivity (which is represented by a logistic equation) minus the number killed by hunting.

Next the economics:

$$B_{i,t} = p - c_{i,t}$$

$$\text{If } B_{i,t} > 0, h_{i,t} = 1, \text{ else } h_{i,t} = 0$$

$$H_t = \sum_i [h_{i,t}]$$

An individual hunter's profits are the price per kill minus the cost per kill. If the profit is greater than zero, they go hunting, otherwise they don't. The total harvest on the prey population is the sum of the number of hunters hunting (Σ means sum), because each hunter kills one animal.

This economic model is particularly simple because we have assumed that each hunter makes just one kill in the time period. The equations for the standard bio-economic model, in which hunter effort is included as a variable, the Schaefer model, are in Clark (1990) and many other fisheries textbooks.

Finally, the bio-economic component, where the population size and the economics interact—this is through the hunter costs. We include some individual variability between hunters, which may be due to differences in their opportunity costs or skill (though note that we don't track individual hunters, as the variation is uncorrelated between years and individuals). To keep things simple, we assume that hunter costs are normally distributed around a mean value, and that this mean value is related to the size of the prey population.

$$C_t = a - bN_t$$
$$c_{i,t} = C_t + \varepsilon(0, s)$$

These equations say that the average cost of killing one animal is a linear function of the population size (the higher the population size the lower the cost), where a and b are constants. The cost that an individual hunter faces is the average cost plus a random component which is taken from a normal distribution with a mean of zero and standard deviation s.

5.3.3 Coding the model

This is the step that people find the biggest barrier to modelling, but with confidence and a methodical approach it is not actually intellectually difficult—the hard thinking should have been done in the previous step. Here's how to go about learning to code a model for the first time:

5.3.3.1 Step 1—Developing a spreadsheet model

Find a paper that has used a model similar to the one that you would like to produce. Ideally the model will be relatively simple, and the data and functional forms used will be well documented and clearly laid out. Replicate the model equations in a spreadsheet. You probably won't be able to do a full facsimile of the model, but you should be able to produce a deterministic non-dynamic version. Make sure that you are generating the same results as the authors—if the inputs are identical you should get identical results, and not just similar ones. For example, you could try re-creating in Figure 5.3 or Figure 1.1. If you don't, dig away at the problem until you are sure you know why not. Depressingly, often a mistake in the published paper is to blame—but be sure!

You can get a fair way with a spreadsheet model, especially if you use macros and add-ins. But the limits come with multiple simulations to carry out sensitivity analyses, particularly when the model needs to be run for a number of years. When this happens, you need to use a proper programming language.

Tips for effective spreadsheet use:

- Declare all your variables at the top of the spreadsheet, so that you can change the numbers easily. No formula should include a number, only cell references. This is a useful discipline for when you start to programme.
- Use relative cell referencing: In Excel, = A4 means that cell A1 will be referenced always, to wherever you copy the formula. = $A1 means that as the formula is copied, the column will stay constant but the row will change, = A$1 keeps the row constant but lets the column change, = A1 means the formula is completely relative. The key F4 lets you switch between these.
- Plot up relationships between variables (for example, between the population size and the mean hunter cost, or between total deer population size and the number of offspring produced), and check that they are as you expect.

Deer model—spreadsheet version

We do the model with constant survival and hunting mortality rates, and density-dependent fecundity as in the equations above. The initial parameter values are:

Survival rates:	Fecundity rate:
$S_1 = 0.6$	$a = 0.6$
$S_2 = 0.9$	$b = 0.02$
$S_A = 0.8$	Harvest rate:
	$H = 0.05$

The spreadsheet should comprise a series of columns, one for each year, in which the rows are:

Row 1 (juveniles): Adults last year multiplied by the fecundity function above
Row 2 (sub-adults): Juveniles last year multiplied by S_1
Row 3 (adults): Sub-adults last year multiplied by S_2 plus adults last year multiplied by S_A minus adults last year multiplied by H.

This set of fecundity values gives a fecundity rate that declines fairly linearly from about 0.65 offspring per adult at very low population sizes to about 0.2 offspring per adult at a population size of 100. Starting from any population size and structure, the model is run until the stable population size and structure is reached, which in this case is a population of 68 individuals of which the majority are adults:

Stable age structure:
Age 1: 18
Age 2: 11
Adult: 39

It is then possible to play around with the fecundity function, plotting up its shape each time you change a or b, and with the values of the other parameters. How much hunting is required to bring the population below two individuals? Can you alter the model so that you have discrete individuals rather than fractions of an individual, and which would be more realistic? What effect does this have? Can you alter it to make survival into a coin-tossing exercise rather than a proportion, so that for each individual, you use a random number to decide whether it lives or dies?

Hunter model—spreadsheet version

In the first column we put the values of the constants in our model. Let's have as an example:

Price (p)	105
Carrying capacity (K)	1000
Intrinsic rate of increase (r_{max})	0.2
Hunter cost intercept (a)	200
Hunter cost slope (b)	0.2
Cost standard deviation (s)	10

Then there should be a column for each year up to, say, 30 years, with a row each for the population size, N_t, (starting with, say, 500 individuals), the productivity, P_t, the average cost, C_t, and the number of animals caught, H_t. To work out H_t for a given year (column), we first need a cost, $c_{i,t}$, row for each of, say, 200 hunters. In Excel, the function for generating a normally distributed random number is = NORMINV(rand(), mean, standard deviation). In our case, the mean is C_t and the standard deviation is s. Then below this, for each hunter decide whether hunting is profitable or not, with a row for each $h_{i,t}$ value [use the formula = IF $(c_{i,t} > p, 0, 1)$], and add up the number killed to get a value for H_t. The final step is to seed the next year (column) with a new population size, which is $N_t + P_t - H_t$. Each time you press F9 you will get a different set of values for your model, as the individual costs are recalculated. You can then play with the model, and see how changing the values of the variables affects the equilibrium population size.

5.3.3.2 Step 2—Programming language

Next you will replicate your validated spreadsheet model in a programming language (Box 5.1). One way to do this is to use the macro function in Excel as a way of learning visual basic, but this is fairly limiting. If you have already learnt R for statistical modelling, then you can carry on using R for this type of modelling too. R is adequate for most simple models, though it struggles with large models such as spatial or individual-based ones. If you have no prior preferences, then use C++, as it is a powerful and generic language. Once you have learnt one programming language you will find others easier to pick up, so the decision is not something to agonise over.

If possible, don't just buy a 'teach yourself C++' book and work through it—it's a very slow and demoralising way to learn. A user manual is a necessity, but as a reference guide, not a tutorial. The best way to learn is to get hold of a simple model that someone else has written in the language that is similar to the one that you want to write (e.g. the one in the Appendix to this chapter). You could ask colleagues or even write off to the authors of the paper you have been using. People are usually happy to share their source code. Go through the model and try to understand what each step means, and in particular, find the equations at the heart of the code. Change them and see what effect it has. Substitute their equations for others (e.g. for the logistic equation). The difficult bits that you absolutely need to

---

Box 5.1 Tips for successful programming.

Model-building is a slow and laborious process that involves great attention to detail. Ensuring that you have a model that is doing what you think it is doing is the longest part of this—but it is vital in order to ensure that there are no nasty surprises later when you come to use the model predictively.

- Use easily distinguishable, short meaningful names for your variables, ideally linked to the notation used in building your conceptual model. This makes the code more readable and so reduces errors.
- Indent your loops (e.g. for time or for multiple simulations), one indent per loop.
- Comment everywhere—notes to yourself and others about what the code is doing.
- Modularise the programme, so you have variable definitions in one file, sub-programmes (units) in another, and only the bare bones of input and output in another. If your code is one long document, it is much harder to debug.
- Declare all your variables up front and NOT inside the code. This is important for spreadsheet models as well. It is virtually impossible to change variables for sensitivity analyses without causing errors unless you have just one place in which the variable is changed.
- Debug the model carefully and deliberately, starting inside each function in turn and then moving out. There will be mistakes, so aim to detect them.
- In debugging runs, start with the deterministic model, and use extremes for each variable, as this is where bugs often occur. What happens when a variable (such as population size or travel cost) is zero? Does the model blow up?
- At each stage, predict to yourself what the result of a particular set of inputs will be—particularly using extreme values, and doing it a function at a time then joining them together. Run the model and see if the prediction is borne out. If it isn't, why not?
- Never let it rest if you can't explain the output—always keep digging until you can.

get right are the routines for starting and finishing the programme, reading in data and outputting results.

Check that you can exactly replicate your results from the spreadsheet model in your programme. When you can do this, you have successfully built a model, and can now alter it to fit your particular system, and go on to use it for sustainability analyses.

It isn't possible to give a detailed tutorial here on how to produce a model in any particular programming language, but in the Appendix to this chapter we give some C code for the hunter model, to give you a feel for what a programme looks like.

5.3.4 Model exploration

Once the model is working and has been fully tested, it can be explored both for its sensitivity to changes in parameter values, and to check that its output bear some relationship to the real world. Models consist of a set of **input parameters** (constants and initial values, such as the parameters of the density-dependent fecundity equation and the initial age structure in the deer model) and a set of **output parameters** (such as population size over time in both models), which are the parameters we are interested in predicting. The input and output parameters are related via the equations that we have specified—and hence a third component of the model that also needs testing is these structural assumptions (Section 5.3.6).

It is vital to explore the model's robustness to variation in input parameter values. This variation could be due to **observation error** (uncertainty about the true value of the parameter due to the data collection process) or to **process error** (environmental variation), and both are important to test for. Sensitivity to process error can tell us which component of the system may be most useful to target in a conservation intervention, while sensitivity to observation error tells us which parameters we need to collect more data on in order to make robust predictions. In practice the two types of error are often confounded, both in real life and in model testing. People are often not explicit about which they are considering in their model exploration, and one can have sympathy for this, given how difficult it is to obtain data on observation error. One approach to this is to build an explicit model of the observation process, which we discuss in Section 7.5.2.

There are two main types of model exploration, elasticity analysis and sensitivity analysis. These two approaches are closely related. In both cases, the idea is to vary input parameters and monitor the effect on output parameters. **Elasticity analysis** calculates the slope of the relationship between an input parameter (say adult survival) and an output parameter (say population growth rate) at a given value of the input parameter. **Sensitivity analysis** involves evaluating the effect on the output parameter of varying the input parameter over a range of values. Before describing these analyses, a note on model output parameters, as it is not always obvious what these should be. For example, ecological studies and population viability analyses (PVAs, Section 5.4.1.1) often look at population growth rate. Although this can be useful to show whether the population is likely to decline or increase significantly under a given set of circumstances, in a stable population at equilibrium the population growth rate is zero. This means that growth rate on its own doesn't tell the whole story when our focus is sustainability. **Population size** is also useful, particularly when it's related to a biological reference point like MSY (Section 4.2.1). But in a bio-economic system there may be other parameters that are relevant to sustainability; perhaps the number of hunters, their catch per unit effort, the profits made, or the number of animals killed.

5.3.4.1 Carrying out elasticity analyses

An elasticity analysis in a simulation model involves varying each parameter in turn by a small amount (typically 1–10%) above and below its baseline value and recording the percentage change in the output parameter of interest. For example, we may look at the percentage change in the population size at equilibrium with a 1% change in each of the survival rates in the deer model. This tells us which parameters have the biggest proportional impact on our output parameter. Elasticities can be calculated analytically if the model is simple enough, which can be very useful (Caswell 2001). They are also scale-independent, so they can be compared between different vital rates and between species.

There has been much debate about how useful elasticity analysis is in conservation and ecology (e.g. Benton and Grant 1999; Mills *et al.* 1999). The general message is that elasticity analysis is a useful component of model exploration, but is not adequate alone. Instead we also have to carry out sensitivity analyses across the range of values that is observed in nature—this gives us a better feel for the likely effects of realistic conservation interventions, and of variation in the environment. For example, elasticity analysis for many age-structured populations shows that the population growth rate is most sensitive to adult survival and not particularly sensitive to juvenile survival. But in ungulates, adult survival varies very little, while juvenile survival is often highly dependent on environmental conditions. These observations are linked, because natural selection favours canalisation of traits for which variation has a strong effect on population growth rate (Eberhardt 2002; Gaillard and Yoccoz 2003). In practical terms, this means that if adult survival is already 98%, then even though increasing it to 99% might substantially improve a population's chances of not going extinct, it may be impossible to achieve. Even if variation in turtle egg survival is not a major contributor to population growth rate or size because it is naturally low, it may still be of conservation relevance to explore the effect of a 20% increase in collecting pressure on the population. These kinds of issues can be addressed by sensitivity analysis.

5.3.4.2 Carrying out sensitivity analyses

Sensitivity analyses are particularly appropriate for simulation models, where the strengths of elasticity analyses (scale-independence and analytical tractability) are not so relevant. The first set of tests to do is in turn to vary each of the input parameters across a **range of values**, keeping the other values the same. This range will be based on 'reasonable' values from the literature, but should also include extreme

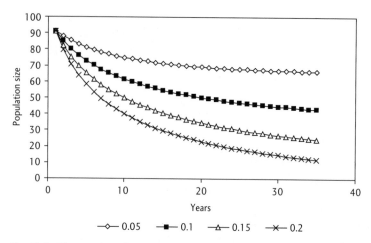

Fig. 5.3 The results of varying the hunting mortality in the deer model, shown for 4 hunting mortality rates, 0.05–0.2. The population starts at the unhunted equilibrium. Tip for plotting graphs—keep them simple and clear, with only a few lines so that they are legible in black and white.

values. For example, in the deer model, the harvest rate should vary between zero and a rate high enough to cause extinction of the population. This ensures that you observe model behaviour over a wide range of possible scenarios, giving you a good intuitive understanding of the model. Plot the results up (Figure 5.3).

Already these analyses can tell you something about the sustainability of the system—how it responds to hunting, for example, or to price changes in the hunter model. They also confirm that you have properly tested your model, because you should be able to predict the shapes of the curves you produce, based on experience and the shape of the underlying functions.

However, in reality parameters do not vary independently, one at a time. In fact it is likely that variation is correlated in space and time—if there is a bad winter it is likely to affect all age classes and all the populations in the region, and to impact on both survival and fecundity. One simple approach to this issue is to simulate baseline, best case and worst case scenarios using the feasible range of values that each parameter can take. The baseline scenario represents your best guess at the parameter values, the best case has all the key input parameters set at the end of the plausible range that gives the best outcome for your parameter of interest, and the worst case has the input parameters set at the other end of the range. This is useful for exploring the effects of our ignorance about parameter values (i.e. observation error) on model outcome.

A more sophisticated alternative is to investigate the sensitivity of the model to variation in a single parameter (say adult survival) in the context of the other parameters, by picking values for each parameter independently at random from their plausible range (Kremer 1983) and repeating this randomisation many times.

Box 5.2 Modelling the effects of harvesting on green turtles.

Chaloupka (2002) gives a helpfully detailed account of the modelling process and parameter values used to develop a stochastic stage-structured simulation model of green turtle (*Chelonia mydas*, Figure 5.4) population dynamics for the Great Barrier Reef in Australia. The modelling process is fully laid out including the conceptual model, a deterministic model, and a simulation in which parameter values are picked from a distribution. Finally the effects of harvesting are included, showing that even low levels of harvesting are likely to lead to population decline. There have been several previous models of turtle population dynamics using elasticity analyses, which suggest that fertility is not an important determinant of population growth (e.g. Crouse *et al.* 1987). These studies have cast doubt on nest protection and head-starting of hatchlings as conservation tools and suggested instead that conservation effort should focus on adult survival at sea. The problem with this is that adult survival is difficult to influence, being mostly to do with interactions with fisheries, while nest protection and head-starting are more feasible. However, using this more realistic approach to evaluating the contribution of particular vital rates to population growth, based on regression analyses over a plausible range of values, Chaloupka showed that fertility is actually a key determinant of population growth.

Fig. 5.4 Green turtle nesting. © Nick Hill.

Then the value of an input parameter in a given year can be related to the output parameter over the full range of all the parameter values, using a multiple regression. This approach is becoming increasingly widely used nowadays (e.g. Wisdom *et al* 2000; Fieberg and Jenkins 2005; Katzner *et al*. 2006, Box 5.2).

Although the multiple regression approach is powerful, there are some caveats. Firstly, it is no substitute for getting a full understanding of the model through the more simple model exploration described above. It is tempting when using this technique to just let your model produce 'data' which are then amenable to statistical analysis—but this turns your model into a **black box**, which is a highly dangerous thing to do, because you can't then know when your model is not working as you think it is, through a programming bug or an unrealistic assumption that you have made in developing the conceptual model. Models are extreme simplifications of reality, which are there to help you explore the consequences of your assumptions about how the world works. If the model is so complex that you don't have an intuitive feel for the mechanisms that are operating, then it stops being a useful tool and becomes a liability. Even complex models can be broken down into manageable sub-models which can then be explored and understood.

Second, it is important to follow normal **statistical good practice** when analysing model outputs—ensuring that the relationships between the model inputs and outputs are linear before testing them using a linear model, for example.

5.3.5 Incorporating uncertainty

So far we have varied parameter values in order to carry out sensitivity analyses. But models can also be built explicitly to incorporate environmental variation that changes demographic rates from year to year, or observation uncertainty in how managers perceive the system that they are trying to manage. Model uncertainty (our confidence in the model itself) can be addressed using similar methods to those discussed in Section 4.4.1 with respect to statistical models, and is also ideally suited to Bayesian modelling approaches (Section 5.4.4). We discuss observation uncertainty in Section 7.5.2. Here we discuss ways to incorporate **process uncertainty** into models.

There is a trade-off to be made in deciding how much variability to incorporate into a model. The real world is variable. There are large-scale events which can knock a system out of equilibrium, such as a major episode of poaching or a severe drought or disease epidemic. There is also the normal variation in weather and chance events such as accidents. The modeller needs to decide how much of this variability to incorporate into a model, in order to ensure that the results are realistic, but that the model doesn't become so messy and complex that its heuristic value is lost.

Demographic variation (for example, in the sex of individual offspring or the fate of particular individuals) is not usually important at population sizes of about 200 or above, but for small populations it is important to incorporate it. For example, in our deer model we used an adult survival rate of 0.8. With 201 adults in year t, this would produce 160.8 adults in year $t+1$, which we can comfortably round to 161. With 21 adults, it may really matter whether there are 16 or 17 individuals next year. The easiest (though not the most elegant) way to incorporate demographic variation into a simulation model is by simulated coin-tossing. For each of the 21 individuals in the adult age-class, generate a random number between 0 and 1. If the random number is less than or equal to 0.8, the individual survives. If it is greater than 0.8, it dies. Note that this procedure doesn't make the model individual-based, because you don't track individuals from class to class over time.

Environmental variation can be important at any population size. The relevance of environmental variation depends on the relative magnitude of the mean and variance of the parameter concerned—if the variance is substantial in relation to the mean (say, as a rough guide, more than 10%), it should be included. Environmental variation can be incorporated in two main ways:

- If the main issue is not day-to-day variation but occasional **catastrophes** (e.g. periodic fires that completely transform plant demographic rates, Morris *et al.* 2006), you can develop several matrices, for example one for good years and one for catastrophe years, and then use one or other each year (for example, if the chance of a catastrophic winter is 0.1, then if your random number is ≤ 0.1 the catastrophe matrix is used).
- For more usual types of environmental variation, one approach is to develop the type of model that is used for regression-based sensitivity analyses, in which a particular year's vital rate is picked from a **distribution**. This could be a uniform distribution (in which each value within the feasible range has the same chance of being picked) or a more complex distribution such as a Normal distribution (see Appendix for code to generate a Normal distribution). If there are data showing that environmental variation leads to correlated variation in vital rates (i.e. in bad years, survival is poor in all age classes and reproduction is also low, and the opposite in good years; Ezard *et al.* 2006), then you can model these correlations explicitly. A simple way to do this is to use correlated Normal distributions. This was done in the example deer model (Milner-Gulland *et al.* 2004). For two correlated vital rates A and B with mean values x and variances s, a stochastic value, ρ, for vital rate A is first drawn from a Normal distribution by using a Normal deviate, z (this is a random number that is picked from a Normal distribution of mean zero and standard deviation 1; see the Appendix for some code to calculate one):

$$\rho_A = x_A + s_A z_A$$

If r is the correlation between the vital rates, a random draw for vital rate B is then given by:

$$\rho_B = x_B + s_B \left(z_B \sqrt{1 - r^2} + z_A r \right)$$

Stochastic models can become quite complex and challenging to programme. However, it is important to include variability in the model if populations are small or variation is large. This is because, for non-linear systems like biological populations, stochastic models do not produce the same mean outputs as deterministic models. A deterministic model overestimates the mean population size in a stochastic system. This is due to a mathematical phenomenon called the lack of

certainty equivalence of stochastic models. There is also an interaction between hunting mortality and environmental variation, such that in general, the higher the hunting mortality the higher the variability in population size, and the further below the deterministically predicted population size the real population size will be—a stochastic model can capture this effect (Hsieh *et al.* 2006). It's also important to consider the degree to which environmental variation is correlated in space and time, as this can have a major impact on the sustainability of harvesting (Jonzen *et al.* 2002).

5.3.5.1 *Practicalities of using stochastic models*

Even if you have decided that a stochastic model is appropriate for your system, it is worth starting model exploration with a deterministic model, so that variation in parameter values doesn't obscure your investigations of model behaviour. When you carry out the full stochastic analyses, you need to ensure that you have run the model for long enough and enough times to produce meaningful results. You need to run the model for long enough in each simulation to ensure that any **transient behaviour** on the way to equilibrium has run through before you start recording data. If you start the model from a near-equilibrium point, the transient period will be shorter. The number of times you need to run the model should be enough that you are getting robust estimates of your output parameters—several hundred simulations will be necessary for most systems you will be modelling.

Which distribution to choose to represent the variation in a parameter depends on the data available and *a priori* theoretical considerations. If the species is poorly known you may only be able to specify a range of values within which the parameter is likely to fall (i.e. use a uniform distribution), while if there are more data you might be able to fit a statistical distribution to the data. Previous studies on similar species may have fitted distributions to data, and you could then use the same distribution. People tend to use the **Normal distribution** as a default. The main theoretical issue is that the variation in the parameter in question should be likely to be Normal or near-Normal, and the main programming issue is to ensure that the value that you get when you pick from the distribution is within the bounds of the parameter in nature. For example, values in a Normal distribution may fall below zero, which is biologically impossible for parameters such as survival or fecundity, while parameters such as survival can't exceed 1. A crude way to deal with this is to have a line in the code saying that if the parameter exceeds 1, it is set to 1. This is OK if the situation is rare, but otherwise, you should consider other distributions. Hilborn and Mangel (1997) have useful pseudocode for generating some of the common distributions.

5.3.5.2 *What can I do if my data are really poor?*

One of the main reasons people give for not developing a model is that their data are too poor, or they don't have an estimate for one of the parameter values. However, models are particularly useful in these situations, because they allow you to explore the consequences of assumptions about the values of given parameters or shapes of functional forms, and to highlight areas of greatest uncertainty where research effort should be targeted. Models are a logical expression of the consequences of our assumptions about system dynamics; we all have implicit models of the system in our heads anyway, and formalising them allows transparency in discussing conservation options. Some ways of obtaining an estimate for a parameter on which the data are poor or unavailable are:

- Use the **literature**, both to obtain estimates for closely related species and to get *a priori* expectations from theory. For example, adult survival can be hard to measure (Section 2.4.2), but we would expect adult female survival in ungulates to be around 0.8–0.9 (Eberhardt 2002).
- **Allometry** is a useful tool for obtaining estimates for parameters which are linked to body size, such as carrying capacity and the intrinsic rate of increase (Peters 1983, Section 2.4.1, Figure 2.10). Although allometric relationships are not always particularly strong, and the differences between a species' predicted and actual values may be biologically important, they at least give a ball-park figure.
- Use your **model** to test out the implications of a particular assumption. This is particularly useful if we have data on all but one parameter. For example, if we have estimates for everything except adult survival and we put in a value of 0.8, what do the age structure, longevity and equilibrium population size turn out to be? Are they realistic values, and are they in line with observations in your population and/or others?

If your model is still parameterised using highly uncertain data and using default functional forms (such as linear density dependence), you may rightly be uncomfortable about using it for management purposes. But it can still be worth building the model, because the exercise itself exposes our ignorance about the system. Here are some useful things to do with a model such as this:

- Vary the parameter values over the full range of possibilities, and look for **qualitative shifts** in model behaviour. Under which circumstances does the population decline to extinction? In which areas of parameter space is model behaviour stable rather than fluctuating or cycling? You can then look to see if your best guess parameter values place the system in a 'safe' area of parameter space, or near a boundary with an unsafe area. For example, how precautionary should a harvest strategy be to ensure that, given our best understanding of the system, we keep population size within safe limits?
- **Invert the questions** you ask. Instead of asking what level of hunting mortality is safe given the fecundity rate you assume, you could ask how high the population's fecundity rate needs to be in order to support sustainable hunting with a given quota. What is the minimum level of spillover of harvestable fish from a Marine Protected Area that will provide a sustainable livelihood for local fishers? And then you can ask how likely it is that the fish dispersal parameter is at this level.

- Use your **sensitivity analyses** as policy outputs. If you find that the model is very sensitive to the value of the lag in harvester effort as it responds to changes in prey population size, and that there is also no information about this parameter, then you have made a useful contribution to the research agenda.

5.3.6 Validation

You may think that your model is a good representation of reality, but you need evidence to back this up if you expect its outputs to be used in conservation. Validation of the model against independent data is a crucial component of the modelling process. In an ideal world, models should be developed with one set of data and then tested with another set, to demonstrate that the model is able to predict datasets other than the one which is used for model development. For example, when looking at the relationship between the presence or absence of a protected species and various factors such as habitat type and hunting level, you can use half your data to develop a model of the relationship between the parameters, and the other half to test the ability of the model to predict correctly (Carroll *et al.* 1999). Or if there is a long time-series of fish catches, half the time-series can be used to develop a model of stock dynamics, which can then be used to predict the other half of the time-series, and the prediction compared to reality.

If you are comparing between models with different **structural assumptions**, validation data can help you to decide which model best predicts the data, and so which one the weight of evidence suggests is most likely to be true. This can be done using the AIC to measure the deviance between the output data that each model produces and the real data (Section 4.4.1). Harrison *et al.* (2006) used this approach to distinguish between a set of models with different assumptions about grey seal movement behaviour.

There are very few conservation models in the literature that are robustly validated, which is a major concern. Usually there are good reasons for not validating properly. There may be few spatial replicates and time-series are short. Often there is little opportunity to learn about the system through adaptive management due to the species being highly endangered (Section 7.5.2.6). But partial validation is still possible, and must be attempted if people are to take your model seriously as a tool for management. Some suggestions are:

- Although it is not true validation, it can still be helpful to show that the model is **internally consistent**, i.e. it is properly predicting the data from which it came. Often there are parts of a dataset that are not used directly as model inputs, which can be tested against model outputs. For example, in the deer model, it may be that there is a time-series of population sizes or age structures, but the model is only developed using the data on survival and fecundity. It is then possible to initialise the model with the age structure and population size that was observed in the first year of the time-series and see if it is able to produce the age structure and population size observed in the last year (Figure 5.5).

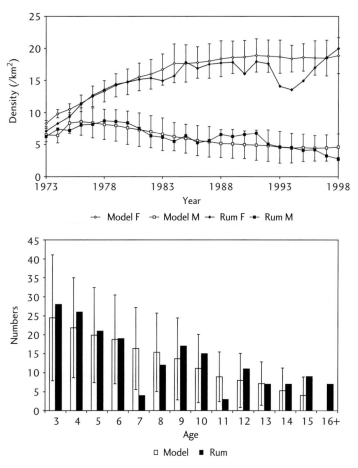

Fig. 5.5 Model check for the deer model in Milner-Gulland *et al.* (2004). This is not true validation, because the survival and fecundity rate data that were used to develop the model are inextricably linked with the age-sex structure of the deer. The model was parameterised with data on red deer demography on the island of Rum. Then it was run starting from the observed population size and structure in 1973, until the end of the data period in 1998. (a) The mean and confidence interval over 50 simulations are shown for the density of mature stags (Model M) and hinds (Model F), together with the equivalent actual densities on Rum (Rum M, Rum F). (b) Mean adult hind population structure (with 95% confidence intervals) for the model in 1998, compared to the actual population structure.

- It is often possible to validate the model predictions quantitatively against **studies in the literature**. For example, Rowcliffe *et al.* (2003) developed a model of bushmeat hunter behaviour based on optimal foraging theory, and showed for several different hunter groups in different countries that the model could predict their offtake rates based upon the model's input parameters such as an animal population's size and movement behaviour.

- It is also possible to validate the model **qualitatively**, by showing that the broad system behaviour is as has been observed previously in this or other systems. This can be done through generating hypotheses based on the model, which can then be tested. Qualitative approaches can be useful in distinguishing between models with different structural assumptions; do some models produce behaviour that is more in line with what we know about this system than the others?

This qualitative validation can lead into a research agenda, which is one of the strengths of a modelling approach. For example, if you adjust the price per unit of prey in our hunter model example, you will find that the number of hunters and the prey population size adjust automatically such that the average hunter cost becomes the same as the price. This is an outcome of the model rather than an assumption, but it fits the predictions of the theory of open access resource harvesting developed using analytical models (Clark 1990). The prediction could then be tested in the real world by assessing whether or not hunters in open access systems are, on average, making profits over and above their costs, and whether or not an increase in the price hunters gain from their prey leads not to more profits, but to more hunting.

Alvard (1993) took this hypothesis-testing approach when he used data to test whether Amazonian hunters acted as 'ecologically noble savages' conserving their resources, or whether they acted like 'rational economic man', hunting according to some measure of short-term profitability. Both hypotheses predicted that prey populations would be depleted near the village, and that hunters would instead hunt in areas further away where prey were more abundant. But he predicted that if someone with a conservation motive came across a prey item on returning home through a depleted area they would leave it, while a rationally economic hunter would kill it. His empirical work showed that the latter was what occurred. Note that this test was based on a conceptual model, rather than a mathematical one.

5.3.7 Scenario exploration

Scenario exploration is when you can finally use your robust and validated model to address the issues of concern to you, and possibly to make management recommendations. In Chapter 6 we explore various management interventions, such as imposing harvesting quotas, setting aside no-take areas, or increasing the opportunity cost of hunting by providing an alternative livelihood (Box 5.3).

Scenario exploration can include many of the techniques from sensitivity analysis, such as varying a parameter value over a range (such as the harvest rate or the price per prey item). It can also include elements of the analyses of model uncertainty—what if all vital rates are correlated, with good and bad years, for example, or what if there is a biological or anthropogenic Allee effect, such that population growth rates increase with population size at low levels (Stephens *et al.*

Box 5.3 Linking wildlife harvesting to household economics in the Serengeti.

Barrett and Arcese (1998) used a model to examine the relative effectiveness of different interventions for conserving wildebeest in the Serengeti ecosystem. Their model was innovative in not just focussing on the sustainable levels of wildebeest hunting, but including household economics—specifically demand for meat and the wage rate from agriculture, as well as a constraint on the amount of work time available for people to allocate either to hunting or to agriculture. This meant that they could look at the way in which people changed their behaviour dependent on the relative profitability of hunting and farming. Their output parameter was the time until the wildebeest population reached crisis levels, and they examined 27 different scenarios with differing levels of correlation between rainfall, wildebeest survival and agricultural productivity; different amounts of meat given to households by the conservation project; and different human population growth rates. Each scenario was simulated 300 times, and the results were presented as the median and range of the time to crisis.

One result was that by giving people meat, the project might actually worsen the situation for wildebeest, as this effectively increases people's incomes and so increases their demand for meat, which is then met by poaching. They next simulated some possible solutions to the problem, which revolved around reducing the correlation between bad years for wildebeest survival and bad years for agricultural productivity, and hence ensuring that wildebeest were not turned to as an alternative to crops when they were themselves vulnerable. This might include improving cropping practices to conserve water and offering employment in years of crop failure.

2002; Courchamp *et al*. 2006)? It is also useful to examine best case and worst case scenarios, to get a feel for the range of potential outcomes of conservation interventions. Another type of scenario analysis is when particular parameters trend over time, hence ensuring that the system cannot reach equilibrium—human population size increasing, perhaps, or carrying capacity decreasing as habitat is destroyed by logging or agriculture. Perhaps the most famous set of scenario analyses are those employed by the Intergovernmental Panel on Climate Change (IPCC), who have four 'storylines' about the future trajectory of climate change depending on the development path that the world takes (IPCC 2001).

The scenarios you choose will reflect the potential issues that conservationists need to address. Then for each scenario, you can examine the outcome of the proposed conservation actions. For example, how quickly would the population recover if there was a moratorium on hunting, under each scenario? Perhaps under some scenarios, the population will continue to decline regardless of the hunting level, due to the effects of habitat loss. The range of situations you can address is - limited only by your model structure and your imagination. Both sensitivity analyses and scenario analyses can generate huge quantities of data, and presenting these data in a digestible form is an art. The first step is to have a robust metric that captures the key question you wish to ask—for example, in Box 5.3, the metric is the 'time to crisis'. Then you can use tables or graphs to show how the metric varies between scenarios.

5.4 Which model should I use?

The main consideration in choosing a model is **parsimony**. The model should be the simplest possible that allows you to answer the question. This is because models are most effective when they are simple enough to provide an intuitive understanding of the system, and all the unnecessary components are stripped out so that the key drivers of the system can be explored without distraction. Complex models have more parameters contributing to model error (both in terms of debugging and in terms of the uncertainty inherent in parameter estimation).

Deciding what the key drivers of system dynamics are is a matter of having a really good understanding of the system, backed up by a thorough read of the literature to see what people have done in similar systems. For example, is spatial structure a key component of the system? Is it necessary to separate the prey population into age, sex or size classes? Is it important to include stochasticity or not? If you are unsure, a useful way to address the level of complexity required in a model is to start with the more complex version of the model, and then **compare its results** to those generated by a simpler version—is the behaviour of the simpler model qualitatively different to that of the more complex and presumably more realistic model? If not, choose the simpler model. People who have done this kind of testing tend to find that simpler models are more robust and reliable than complex ones. For example, Hilborn and Walters (1987) used a relatively complex age-structured model to generate a dataset, and then used this dataset to test the predictive power of both the same model and a simple lumped population model. The simple model performed better, even though it hadn't been used to generate the data. The simplest approach was also preferred in the competition run by the International Whaling Commission to find a model for whale management—in this case, a statistical model that fitted a curve to the empirical data out-performed more complex mechanistic models (Box 7.7).

5.4.1 Biological models

The best model to use depends both on the biology of the species and on the management options you will be exploring. The main division when modelling populations is between lumped and structured models, but there are several ways in which a model can be structured (Table 5.2). Biological models can also be at the genetic, community and ecosystem levels, but these are less commonly used for modelling the sustainability of hunting.

Table 5.2 Types of population model.

Model type	Use
Lumped parameter	Default when no reason to use more complex models, e.g. logistic growth model.
Age-structured	Long-lived species when biological processes are age-related (e.g. juvenile/prime adult/senescent), e.g. many ungulates.
Sex-structured	When harvesting is sex-biased, e.g. for male trophies, and/or when sex differences are biologically important, e.g. harem-breeders.
Stage-structured	When size (weight or height) or stage (e.g. larva/adult) is more important, biologically or for harvesters, than age, e.g. trees, fish, insects.
Spatially structured	When dispersal is a key biological parameter, e.g. fragmented habitats, meta-populations and/or when management is spatially structured, e.g. marine reserves.
Individual-based	Very small populations or when individuals are sessile or territorial and spatial structure is important (e.g. duikers, plants).

Many models nowadays are structured in several ways. For example, Skonhoft *et al.* (2002) produced a model of chamois hunting that included movement between a protected area and a hunting zone (spatial structure) as well as age and sex structure, because males are preferentially hunted for their horns. They also included a socio-economic component, exploring which parameters had a major effect on the hunting rate that would maximise overall profits.

Spatially structured models are becoming more widely used now, and it is evident that space is a key component of the sustainability of harvesting, both from the biological side and from the socio-economic side. For example, remote areas can provide a refuge for over-exploited species (Ling and Milner-Gulland in press). Dispersal rates between areas are key biological parameters determining the success of spatial harvest controls such as no-take zones (Section 6.4.1)—too high, and the animals disperse out of the protected area and are harvested, too low, and there are few economic benefits to harvesters (Gerber *et al.* 2003).

5.4.1.1 Population viability analyses

Much early conservation modelling, particularly of small populations, and particularly using packages such as Vortex and RAMAS, has been labelled as Population Viability Analysis, or PVA. The distinguishing feature of a PVA, as opposed to any other kind of population model for conservation, is that it aims to estimate the **probability of population extinction** by using a stochastic model.

There has been debate about the usefulness of PVAs as a tool for conservation (Brook *et al.* 2000; Coulson *et al.* 2001), which is relevant to any kind of modelling exercise that aims to provide management recommendations. The main issue with PVAs is whether the extinction probabilities that they predict have any meaning, particularly when the probability of irregular and rare events like very bad weather, disease epidemics or poaching sprees may be very hard to predict despite having a major influence on probability of extinction. The evidence from extensive testing of PVAs in simulated environments suggests that they are useful in ranking management alternatives, even if they can't necessarily accurately predict the probability of extinction itself (McCarthy *et al.* 2003).

The distinguishing feature of a PVA is not the form of the model, but the metric that is used to assess conservation performance—a PVA by definition needs to produce an estimate of extinction risk. This makes them a fairly limited subset of the types of model that are used in assessing the sustainability of use. Partly this is because extinction is not likely to be a probable near-term outcome if a population is large enough for sustainable use, and partly because we take a broader view that includes social and financial aspects of sustainability. Hence, although the literature on PVA is interesting and relevant, it is a relatively small part of the modelling toolkit.

5.4.2 Behavioural models

Behavioural models can be very useful for predicting the behaviour of both prey and hunters. Within behavioural ecology, **optimal foraging theory** provides a range of useful models for predicting prey choice and spatial patterns of foraging, which can be applied just as well to humans as animals (e.g. Alvard 1993; Winterhalder and Smith 2000; Rowcliffe *et al.* 2004). From the economics side, human hunting behaviour can be modelled using similar techniques, so hunter effort varies depending on the availability of prey in a particular location. For example, Moustakas *et al.* (2006) produced a model in which both fish and fishers moved through a grid of cells, with the fish looking for foraging and spawning grounds, and the fishers moving to where the fish were most abundant. Another set of behavioural models comes from **game theory** (Binmore 2007). This predicts how individuals behave when their optimal decision is dependent on the decisions made by others, and is most useful when there are only a small number of individuals involved. For example, Mesterton-Gibbons and Milner-Gulland (1998) used game theory to explore the incentives to monitor others or to poach in a community-based wildlife management project.

Recently, there has been a lot of interest in **agent-based models**, which hold out the prospect of simulating the strategic decision-making of individual hunters. These models are a sub-class of individual-based population models (IBMs, Table 5.2), and are as computationally intensive as all IBMs. They can also be difficult to programme, although there are a number of software packages available (see Resources section). The distinction between agent-based models and other IBMs is that the agents respond to their environment, and can therefore act strategically. In other individual-based models, the individuals just have to be distinguishable, not necessarily responsive. For example, Bousquet *et al.* (2001) produced a spatially explicit model of the interaction between hunters and duikers in Cameroon. Duikers are territorial species, and so the model is spatially explicit, with each duiker occupying a particular cell on a grid, taken from a GIS of the area around the village in Cameroon. The duikers make movement decisions based on the presence of other duikers in the region, while hunters make decisions about where to place their trap networks. It is then possible to overlay a management regime controlling the hunters, in a way reminiscent of the operating model approach to fisheries management (Section 7.5.2.3). Agent-based models are also potentially useful for simulating strategic interactions between people (e.g. poachers and monitors in community conservation schemes) when the system is too complex for standard game theoretic approaches. They have also been used to simulate the multiple institutional levels at which decision-making occurs in the management of natural resources (e.g. Walker and Janssen 2002).

A more standard way of simulating economic behaviour is the **household utility model**, where households or individuals act to maximise their utility in the face of different productive options and a limited amount of available labour; this is the type of model used by Barrett and Arcese (1998; Box 5.3) and Damania *et al.* (2005). Utility is the basic unit of economics, and measures happiness or human welfare. It is often approximated by money and/or consumption, although resources are clearly also valued by people in ways that are difficult or impossible to quantify. A purely financial definition of utility is therefore convenient, but likely to be misleading when non-financial values, such as cultural significance, are important.

5.4.3 Bio-economic models

All models that aim to say something about sustainability need to include both the biological and the human components of the system, and how they interact. Basic models of this type include those founded in logistic population growth and open access harvesting, such as those discussed in Chapter 1. But these are very simplistic, and so are more useful in helping us to understand the broad behaviour of bio-economic systems, rather than for making management recommendations in real systems. In real systems, issues like the lag between a change in the system and the response to it can be critical to system behaviour—for example, hunters take time to change their behaviour if the cost of hunting increases, while prey populations need time to reproduce and grow if hunting pressure is reduced. These lags can lead to cycles in the system's dynamics, at least in theory (Sanchirico and Wilen 1999).

As usual, fisheries modellers are leading the field in producing models that include both human and biological elements in more and more realistic settings. For example, Pelletier and Mahevas (2005) review a huge range of models that have been used to assess the performance of Marine Protected Areas (MPAs) in conserving fish stocks while at the same time providing a sustainable yield for fishers. The models started very simple and highly theoretical, with only two patches and dynamics based on the Schaefer model (Chapter 1). More recently, research has focused on trying to predict how fish with different biology respond to MPAs; what happens if adults rather than larvae disperse, for example, or if fish have particular areas where they go to spawn or forage at particular times of year? On the human side, researchers are exploring the way in which individual fishers respond to restrictions on their activities, and how they cooperate to find fish (Smith and Wilen 2003; Branch *et al.* 2006).

Dynamic bio-economic models can become very complex because of the necessity to include both the human and biological components of the system. Improved computing power means, though, that it is realistic to think of carrying out simulations that go back to first principles, and use simple decision rules by both people and prey, rather than having to solve these models analytically. This would involve building on simple models such as our hunter example above, though models involving large numbers of individuals and great spatial detail can soon become unmanageably large and slow to run on the current generation of personal computers.

5.4.4 Bayesian models

Bayesian modelling approaches are very widely used nowadays, particularly in fisheries managament, although they are also being advocated for conservation and natural resource management more generally (e.g. Wade 2000; Dorazio and Johnson 2003; Ghazoul and McAllister 2003). Ten years ago, McAllister and Kirkwood (1998) gave an excellent and accessible review of the uses of Bayesian models in fisheries, and said that these methods were still relatively inaccessible to most scientists. Nowadays, there are software packages available such as WinBUGS, which make the actual programming of models easier (see Resources section), but the mathematical understanding required is still conceptually difficult. The basic maths is **Bayes' theorem**, which is a standard simple equation in probability theory. However, although this is the engine of Bayesian modelling, it doesn't take the novice far in working out how to actually implement a Bayesian model.

Bayesian probability is an alternative to classical frequentist probability theory. In frequentist theory, you assess the likelihood that a hypothesis is true given your data. For example, based on a sample of 10,000 tosses of a coin I can say that the probability of a coin toss producing heads is almost exactly 0.5. However, if I only tossed the coin twice and produce two heads, I would estimate that the probability of getting heads is 1. I could then use standard statistical techniques to test whether the outcome of my experiment allows me to reject the null hypothesis that I have a fair coin with a probability of 0.5 of getting heads (I clearly couldn't reject it in this case, because my sample size is so small).

Bayesian statistics works the other way round—you assess how likely your data are given a hypothesis. This may seem like a subtle difference, but it allows a very important change in approach—you can test your ideas using more information than just your data. In the example above, my data suggested that the probability of getting heads was 1, but my friend says that the probability is 0.5 based on her reading of the frequentist literature. Given this prior information, I could assign a

prior probability distribution that gives a probability of 0.75 to the coin producing heads 50% of the time, and, say, a 5% probability that it produces heads all the time, with the rest of the probability evenly distributed among other outcomes. I am already doing better than the pure frequentist approach, despite my tiny data set. As new data become available, these distributions can be updated, and the credibility of each hypothesis is then reassessed. For example, after 20 goes, I now have 13 heads and 7 tails, so I can reduce the probability of the coin only producing heads to zero, and increase the weighting of the values in the 0.5–0.7 range.

Bayesian statistics are very useful when giving recommendations for management action under **conditions of uncertainty**. This is exactly the situation we find in the conservation of exploited resources. For example, we may wish to give a prediction about the outcome of a particular management strategy using a model of the dynamics of the harvested population. We can collate any data that may inform us, in whatever form it takes, whether it be expert opinion, quantitative data from the system itself, or data from similar systems. These data are used to produce the prior probability distributions for the model parameters, and to develop different **candidate models** for how the system works. As the models are updated based on data, we obtain new understanding which we use to update our priors into posterior probability distributions—this is **Bayesian updating**, and is where the complicated mathematics is required. We can update both our probability distributions for parameters given a particular model structure, and our weighting of the credibility of each model that we test (e.g. see Box 4.5). Next, we can use these updated models to provide management advice. This advice can be framed in probabilistic terms to reflect the underlying uncertainties—the probability of your management producing the desired outcome, given our model of the state of the world A, is X%, and the likelihood that model A reflects the true state of the world is Y%. Finally we would hope that any management that is implemented produces further new understanding, that can be used to update the model priors next time around (Figure 5.6).

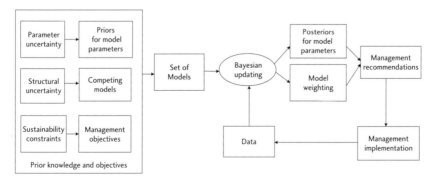

Fig. 5.6 A schematic showing the Bayesian approach to uncertainty in management of natural resources.

This approach goes hand in hand with the adaptive management approach discussed in Chapter 7, and with the approach of confronting models with data promoted by Hilborn and Mangel (1997) and others. It is an extremely powerful and elegant way of dealing with uncertainty, allowing its effects on predictions to be fully incorporated into the modelling process and clearly expressed for managers. However, developing a full Bayesian model is a task for a specialist.

Bayesian modelling is still a very young field in conservation—this is reflected in the fact that at the moment there are many more articles telling us what a useful tool it is than actually using it! Even if you decide not to invest the substantial time and effort required to become a Bayesian modeller, the underlying philosophy can still inform your modelling—you can still explore the effects of uncertainty on your model predictions using the range of techniques discussed in Section 5.3, and can still give management advice that explicitly takes account of uncertainty.

5.4.4.1 Bayesian networks

Bayesian network (or Bayesian Belief Network, BN) models are currently increasing in popularity in conservation and natural resource management (e.g. Marcot *et al.* 2001; Wisdom *et al.* 2002). Their only link to Bayesian statistical models is that they use Bayes' theorem—they are otherwise a completely different type of model. Bayesian networks have a set of nodes connected by directional links, and are usually used to show causative relationships between parameters (Box 5.4).

Bayesian networks are attractive because they are relatively easy to programme and to understand, and because they have accessible and well-documented software available (see below for links). The modelling software available allows sensitivity analyses, exploration of the effects of the addition of new information to the model, and also allows dynamic models that show how the system evolves over time. The graphical presentation can be a very attractive way of presenting results to end-users. However, there are some caveats as well:

- **Cycles** are not allowed, i.e. a node can't influence itself, even indirectly. This causes the model to blow up, but can generally be easily solved by using a dynamic model. For example, animal population size in time $t + 1$ is a function of animal population size in time t, through the action of density dependence.
- The **causative relationships** that are represented in the BN are a major factor in determining the outcome of the model. Just as for any model, the underlying model structure needs to be robust, and if there is doubt about the causative links, alternative versions of the model need to be tested.
- Just as for any off-the-shelf package, the range of **in-built functions** is limited, and may not include what you want to do, particularly as regards model exploration and data entry. The models may also run quite slowly if they are large and complex. However, the packages do allow you to programme your own add-ins, which make them substantially more powerful if you have the technical expertise required.

Box 5.4 Factors affecting sustainability of co-managed fisheries.

In Box 4.6, we discussed the results of Halls *et al.*'s (2002) study in which they looked at factors affecting the sustainability of small-scale co-managed fisheries. They also developed a Bayesian Network model as part of the study. This model can be graphically represented as Figure 5.7:

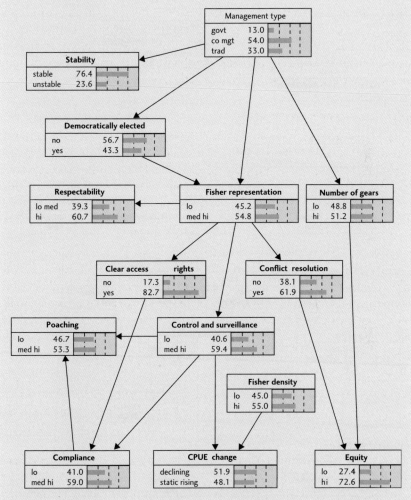

Fig. 5.7 Bayesian network model for the sustainability of co-managed fisheries. The squares represent nodes, each of which has a conditional probability table associated with it, and the arrows are causal links (for example the probability of the fishery being stable is conditional only on management type). The bars for each node represent the overall prevalence of a particular state in the dataset.

They have three outcome nodes that represent aspects of sustainability—compliance, equity and catch per unit effort. These are influenced by a range of factors, each of which is expressed as a set of discrete values (e.g. yes/no or high/low). For example, 54% of the fisheries in the study were co-managed, and the overall probability of a fishery in the study having a declining CPUE was 51.9%. The network represents the chain of causation between the factors.

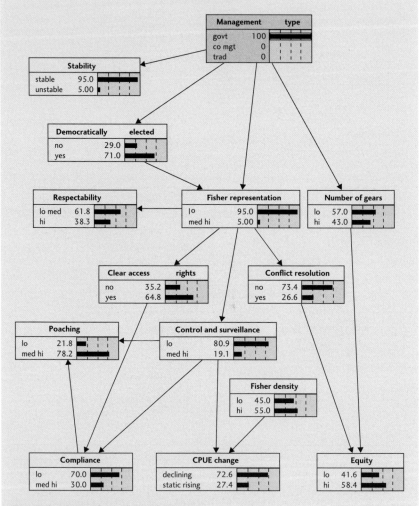

Fig. 5.8 Network model with Government as the management type (showing forward propagation of probabilities—if we know the characteristics of the fishery, we can predict the probability of particular sustainability outcomes).

So CPUE is directly influenced by fisher density and the level of control and surveillance. Behind each of the nodes is a conditional probability table, for example showing the probability that fisher representation on the management body is high as a function of management type and whether the management body is democratically elected. These probabilities can be obtained from any source, such as quantitative data or expert opinion (in this case, it was by using a logistic regression to analyse quantitative data).

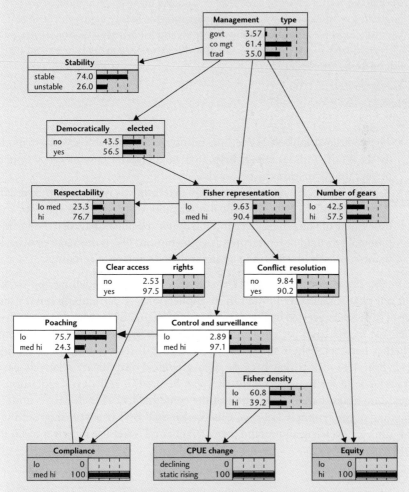

Fig. 5.9 Network model with all three output nodes set at the desirable outcome (showing backward propagation of probabilities—if we know which sustainability outcome we want, we can see what characteristics of the fishery are more or less likely to lead to to this outcome).

The strength of the BN approach is that probabilities can be propagated in both directions through the use of Bayes' theorem. This makes it a very powerful tool for inference. For example, if we wished to know what the effect of the government managing a fishery was, we could set the 'management type' node to 'government', and see how the causation propagated forward through the network (Figure 5.8): The probability of having a declining CPUE has now risen substantially, to 72.6%.

We could alternatively ask what would be required for sustainability, i.e. for all three of the output nodes to have positive values? This again gives some clear and interesting answers about what way of structuring the fishery management works best—for example although clear access rights and fisher representation on management boards are essential, it makes little difference how many fishers there are fishing in the area (Figure 5.9).

Source: Halls *et al.* (2002).

- The model may look attractive, but end-users will not be able actually to modify the model without **expert help**, so it would not be a useful legacy from a project simply to leave the BN model behind.
- Bayesian networks are also surprisingly **deterministic** compared to Bayesian statistical models. The conditional probability tables give one value for the probability of outcome A given input B, but in a standard Bayesian model this probability would be represented as a distribution. This is the major drawback of the approach in terms of its ability truly to represent uncertainty.

In the end, Bayesian Networks are another potential modelling approach, but one that is relatively limited in its applications. A particularly interesting use of BNs was suggested by Hammond and O'Brien (2001), who demonstrated their use as the basis for consensus-building in a hypothetical dispute about whether haddock fishers should be allowed to cull seals to stop them eating their fish. The idea is that because their graphical outputs are relatively easy to understand for non-specialists, they can be used in real-time negotiations, providing a transparent illustration of the consequences of each party's understanding of the system. Cain *et al.* (2003) also used BNs in a workshop setting, this time for deciding about land management and water use in a river basin in Sri Lanka.

5.5 Moving from modelling to action

Predictive models are an incredibly useful and powerful tool for the exploration of conservation options, particularly when empirical data are scarce and hard to come by. They are an integral part of the process of adaptive management, as discussed in Section 7.5. Conservationists have been very reluctant to use models in real-world management in the same way that fisheries scientists have, and so there are very few examples outside fisheries science where a true interaction between modelling and management can be demonstrated (see Box 7.8 for a rare exception). This may be due to the non-quantitative background of many conservationists, the less severe financial consequences of decisions about hunting quotas for species of conservation concern, and the perceived complexity and site-specific nature of human-wildlife interactions at the subsistence level. There is also caution about the defensibility of producing models of human behaviour when profit is not the main motive for hunting behaviour. However, we have hopefully convinced you that models are an integral component of management, and that the mystique surrounding modelling as a skill is unwarranted.

However, it's worth sounding a note of caution. If you have been thorough in your model conceptualisation and exploration phases, you will have a very good understanding of its behaviour and limitations, and you will be able to express these limitations through your sensitivity analyses and scenario explorations. But the most sophisticated model is still only as good as its inputs—**Garbage In Garbage Out**. Our experience is that when models are used to address policy questions, it is the modeller who is the most sceptical of the recommendations arising, while non-modellers who feel that model results support their viewpoint tend to seize on the simple headline figures and ignore the many pages of caveats that surround them. Be very careful, therefore, about how you present your results. Tools from **decision analysis** can be very helpful in presenting results in a way that fairly represents the uncertainty that surrounds them (Section 7.5).

Finally, it can be very easy to get caught up in the virtual world, and think of the model as an end in itself, rather than a tool for analysis. It can be fascinating to explore the mathematical behaviour of models, and some parts of theoretical ecology are very far removed from empirical reality. However, conservation is an applied subject in which we are trying to implement practical solutions in the real world. So it is particularly important to have a very strong sense of the relationship between your model and reality, and to be clear how it is going to contribute to improving practical action on the ground.

5.6 Resources

5.6.1 Websites

Off-the-shelf packages:

RAMAS: http://www.ramas.com/software.htm. Particularly strong on spatially explicit models, also has a stage-structured model.

Vortex: http://www.vortex9.org/vortex.html The original PVA modelling package, quite a strong emphasis on genetic effects and small populations.

Forestry software list: http://dmoz.org/Business/Agriculture_and_Forestry/Forestry/Software/

Some relevant fisheries software, including ParFish: http://www.fmsp.org.uk/Software.htm

Bio-economic spatial modelling for Marine Protected Areas: http://www.ifremer.fr/isis-fish/objectivesen.php

ULM free software for structured population models: http://www.snv.jussieu. fr/Bio/ulm/ulm.html

BUGS, including WinBUGS and links to other sites on Bayesian methods: http://www.mrc-bsu.cam.ac.uk/bugs/welcome.shtml

Agent-based modelling software:

CORMAS (specifically targetted at natural resources): http://cormas.cirad.fr/indexeng.htm

SWARM (more generic, and the page has comparisons of several software packages): http://www.swarm.org/wiki/ABM_Resources

Bayesian Network software:

Netica: www.norsys.com
Hugin: www.hugin.com
Simple online tutorial: http://www.dcs.qmw.ac.uk/%7Enorman/BBNs/BBNs.htm
Comparison of packages: http://www.cs.ubc.ca/~murphyk/Bayes/bnsoft.html

5.6.2 Textbooks

Akcakaya, H.R., Burgman, M., Kindvall, O., Sjogren-Gulve, P., Hatfield, J., and McCarthy, M. (2004). *Species Conservation and Management: Case Studies.* Oxford University Press, Oxford. Useful compendium of matrix population models with a CD of the models themselves, for use with RAMAS software.

Binmore, K. (2007). *Playing for Real: Game Theory.* Oxford University Press, Oxford. Binmore's 1991 textbook Fun and Games was a very good introduction to game theory. This is an updated version, about to come out.

Caswell, H. (2001). *Matrix Population Models: Construction, Analysis and Interpretation*, Second Edition. Sinauer Sunderland, MA. The bible for age/stage-structured modellers—but it is mathematically heavy, and doesn't have the relevant examples of Getz and Haight.

Cowell, R.G., Dawid, A.P., Lauritzen, S.L., and Spiegelhalter, D.J. (1999). *Probabilistic Networks and Expert Systems.* Springer-Verlag, Berlin-Heidelberg, New York. Very well-regarded textbook on Bayesian Networks.

Donovan, T.M., and Welden. C.W. (2002). *Spreadsheet Exercises in Conservation Biology and Landscape Ecology.* Sinauer Associates Sunderland, MA. A step-by-step guide to implementing basic models in Excel.

Getz, W.M., and Haight, R.G. (1989). *Population Harvesting: Demographic Models of Fish, Forest, and Animal Resources.* Monographs in Population Biology 27, Princeton University Press, New York. Our favourite book on the theory behind harvesting models. Particularly good on age/stage-structured models, and lots of relevant examples.

Haddon, M. (2001). *Modelling and Quantitative Methods in Fisheries.* Chapman and Hall Boca Raton, Florida. Useful introduction to fisheries modelling techniques, with a good section on bootstrapping.

Hilborn, R., and Mangel, M. (1997). *The Ecological Detective: Confronting Models with Data.* Monographs in Population Biology 28, Princeton University Press, New York. A must-have for all who wish to carry out quantitative analyses based on empirical data. A landmark work.

Korb, K.B., and Nicholson, A.E. (2004). *Bayesian Artificial Intelligence.* Chapman and Hall Boca Raton, Florida. A useful introduction to Bayesian Network modelling.

Mangel, M. (2006). *The Theoretical Biologist's Toolbox: Quantitative Methods for Ecology and Evolutionary Biology.* Cambridge University Press, UK. Packs a huge amount in—including a chapter on fisheries models. Quite advanced mathematically.

Morris, W.F., and Doak, D.F. (2002). *Quantitative Conservation Biology: Theory and Practice of Population Viability Analysis.* Sinauer, Sunderland, MA. A comprehensive tool-box of techniques for stochastic models.

Appendix

This contains some C code for a simple bio-economic model, our hunter example. Explanatory comments are indicated by //. Because this is a very simple programme, we haven't modularised it, but in a more complex programme you would put the functions and the variable definitions into their own unit files to avoid cluttering up the main file. In order to make this model work, all you would need to do is to import this text into a suitable C compiler and run it. This model was written in C within the Visual C++ environment. Not all C compilers are quite the same (they don't all have the built-in units to support standard functions). If you have trouble running the code, it is most likely to be due to the use of a different compiler.

```
//hunter program, to simulate village hunting

//these are built-in units that support standard functions
#include <stdio.h>
#include <stdlib.h>
#include <math.h>
#include <time.h>


//start program
void main(void)
{
//define constants
#define K 1000            //carrying capacity
#define r 0.2             //intrinsic rate of increase
#define P 105             //price
#define a 200             //const for cost calc
#define b 0.2             //const for cost calc
#define s 10              //cost SD
#define N1 500            //starting population size
#define Years 30          //run the model for 30 years
#define Hunters 200       //200 hunters

//define variables
float Pop[Years],Harvest[Years];    //population, harvest
float Prod,PrF,Cost,IndC,zval,B;    //productivity,costs,
                                    //  random number,profit

int i,t;                            //integers - hunter, year
```

```
FILE *outfile;                          //name of the file for output

float z_calc(void);                     //initialises normal
                                          distribution routine
float round(float PrF);                 //initialises rounding
                                          routine
srand((unsigned)time(NULL));            //initialises random number
                                          generator

outfile = fopen("output.out","w");      //opens output file

//starting population
for (t = 0; t < Years; t++) {Pop[t]=0; Harvest[t]=0;}
Pop[0] = N1;

//*********************** years ***********************

for (t = 0; t < Years-1; t++)
{
PrF = r*Pop[t]*(1-Pop[t]/K);            //logistic productivity
Prod = round(PrF);                      //births must be a whole number
Cost = a - b*Pop[t];                    //mean hunter cost

for (i = 0; i < Hunters; i++)           //hunter decision-making
{
    zval = z_calc();                    //get a random number from
                                          normal dist
    IndC = Cost + zval*s;               //hunter cost
    if (IndC < 0) IndC = 0;             //ensure no negative costs
    B = P - IndC;                       //profitability
    if (B>0) {Harvest[t]++;}            //hunting decision
} //i (hunters)

if (Harvest[t] > Pop[t] +               //can't harvest more than
 Prod) Harvest[t] = Pop[t] + Prod;       is there
Pop[t+1] = Pop[t] + Prod -              //next year's population
 Harvest[t];

printf("%d %.0f %.0f %.0f %.0f\n", t,Pop[t],
 Prod,Harvest[t],Pop[t+1]);            //onscreen output

} //t (years)

//********************* end bits ***********************

for (t = 0; t < Years; t++)
  fprintf(outfile,"%d %.0f %.0f\n",
  t,Pop[t],prod,Harvest[t]);           //output to file

fclose(outfile);
} //end

//********************* functions ***********************
```

```
float z_calc(void)                  //this is a procedure for getting a
                                      value from a normal distribution
{
#define comp 8000
#define pi 3.141592654

float prob,step,LastP;
float incre,cumprob,z_val;
double RNo;

RNo = (double)rand()/(double)RAND_MAX;
 z_val = 0; cumprob = 0;
 incre = 8/(float)comp;
 step = -4;
 while (step <=4)
{
     step = step + incre;
     prob = (1/sqrt(2*pi))*exp(-(step*step)/2);
     LastP = cumprob;
     cumprob = cumprob + (prob*incre);
     if (RNo > LastP && RNo < cumprob)
         {z_val = (step - incre) + incre*((RNo-LastP)/
         (cumprob-LastP));
     step = 4;}
}

return z_val;

} //Zcalc

//************************************************************

float round(float PrdF)             //this is a procedure for rounding a
                                      number
{
float Prdn,diff;
int low,high;

low = (int)PrdF;
high = (int)PrdF+1;

diff = PrdF-low;
if (diff<0.5) Prdn = (float)low;
  else Prdn = (float)high;

return Prdn;

} //round
//************************************************************
```

6

Choosing management approaches

6.1 Scope of the chapter

In Chapters 2–5 we discussed how to assess the sustainability of natural resource use, addressing the biological and socio-economic aspects of data collection, and the use of modelling for analysis. We have highlighted the importance of establishing scientifically the degree to which different factors contribute to biodiversity loss and how they interact, rather than assuming that the most obvious factor is automatically the most influential (Caughley and Gunn 1996). Using the methods outlined in Chapters 2–5, we can obtain information on a range of factors that will allow us to plan management interventions effectively. We give the basic explanations of these factors below, together with illustrative examples in Box 6.1.

- *Issue of concern*. Why is a change in management needed? This is a fundamental question, the answer to which is often taken for granted. But without a statement of the problem and the evidence to back up the assertion that instituting management of resource use is the answer, we could go badly wrong.
- *Resource type*. The basic characteristics of the natural resource. What is its monetary value per unit at the point of harvest and sale, how transportable is it, how perishable is it, and what sort of value does it have to users (both producers and consumers)?
- *Biological characteristics of the resource*. These include abundance, distribution and productivity of the target species, the identity of the target species (one or many taxa?), and by-product mortality to other components of the ecosystem.
- *Harvester characteristics*. Does the harvester live locally to the resource or not? What methods are used for harvest? (What types of gear, seasonal activity, length of hunting trip, is there more than one harvester type operating?) What alternate livelihood activities are open to harvesters? What is the profile of harvesters in comparison to the general population (age, education etc.)? What attitudes do the harvesters have to the resource and their profession?
- *The commodity chain*. The identity, numbers and locations of the actors at each point in the chain. The length and stability of the commodity chain, and the points at which intervention might best be targeted (e.g. bottlenecks).
- *Consumer characteristics*. The location of consumers (local to the resource, in-country but removed e.g. urban, international). The preferences of consumers and substitutability of the resource; the narrower the niche the resource fills, the fewer substitutes it will have. Consumer elasticities of price

and income. Attitudes to the resource, including cultural importance and knowledge of its conservation status.
- *Existing institutional framework.* The ownership and control of the resource and its habitat. Legislation at the national, international and local levels. Traditional and de facto use rights. Cultural mores concerning the resource.
- *Additional threats to the system, potential or actual.* These include both biological and socio-economic threats, such as land use change and habitat loss, aliens and invasives (both immigration of users from outside and alien species), hybridisation and disease.

Box 6.1 Examples of factors influencing management interventions.

The issue of concern—the Tanimbar corella

Just because the use of an endangered population is high profile does not necessarily mean that it is the most significant cause of the population's predicament, nor that stopping it will be the most effective way to ensure population persistence. For example, banning the international commercial trade in the Tanimbar corella (a type of parrot, *Cacatua goffini*) was largely counter-productive (Jepson 2002). This endemic species is listed as near-threatened on the IUCN Red List (IUCN 2006) because of ongoing population declines and its small population size. However, it is locally abundant and the commercial trade provided some compensatory benefits to local people from its crop-raiding activities. The removal of this benefit reduced incentives to conserve the species, and provoked considerable resentment and mistrust of conservationists.

Resource type—High-value products in international trade

Rhino horn is a good example of an attractive export commodity because it has a high value per unit, the units are conveniently sized for transport, it is non-perishable, and has a high cultural value for traditional medicines in the Far East and for dagger handles in Yemen (Martin and Martin 1982). Shark finning, involving the disposal at sea of the rest of the shark (usually still alive), is prevalent because only the fins have a high enough value–volume ratio to make transportation cost-effective (Fig. 6.1). Current initiatives for improving the sustainability of this trade include ensuring that sharks are landed entire, with the fins still attached to the body, thus reducing the economic value of the catch while aiding monitoring (IUCN 2003).

By-product mortality—gaharu harvesting

Examples of bycatch of non-target organisms in commercial fisheries are numerous and well documented (Lewison *et al.* 2004). A forestry example of by-product mortality is the harvesting of gaharu resin, a valuable fragrant wood produced by trees of the genus *Aquilaria* as a response to fungal attack. Traditional harvest methods involve detailed checks to see if an encountered tree contains gaharu before felling; in Indonesia, this practice is much less common and the trade is highly commercialised. Harvesters now cut down virtually all *Aquilaria* trees encountered, healthy or not (Soehartono and Newton 2002).

Fig. 6.1 Fins freshly cut from thresher and requiem sharks at a landing port in eastern Taiwan in 2000. In this longline fishery fins are removed in port because sharks are sold for their meat, however, in many fisheries serving markets with no demand for shark meat, shark fins are cut on deck and the remainder of the carcass discarded at sea. Photo © Shelley Clarke.

Harvester characteristics—attitudes towards law enforcement in fishing communities

One example of how harvester attitudes shape behaviour is from small fishing communities in Norway and Canada (Gezelius 2004). Here, there was qualified acceptance of outside regulation of commercial fishing, and condemnation of those found breaking the law. Illegal fishing for household consumption was seen as harmless and acceptable, while illegal commercial fishing by individuals badly hit by new regulations was felt to be morally ambiguous. Subsistence food fishing was seen as inappropriate for outside regulation, regardless of its legal status. Gezelius argues that these moral attitudes are a fundamental component of Western Christian societies, and may be more widely held.

The commodity chain—the effect of civil conflict on a bushmeat commodity chain

The effect of civil conflict on commodity chains was examined in a study of the bushmeat trade around the Garamba National Park in the Democratic Republic of Congo (de Merode and Cowlishaw 2006). Before the conflict, the commodity chain was controlled by local women. During the conflict, soldiers took over the bushmeat trade, shortening and simplifying the chain. The volume of bushmeat traded increased, as they tried to maximise profits. After the conflict, the original actors re-established their control of the commodity chain and traded volumes returned to pre-conflict levels. This indicates that commodity chains can be resilient to severe disruption.

Consumer characteristics—consumer preferences in Equatorial Guinea

East *et al.* (2005) looked at consumption of and preferences for the main food groups in the city of Bata, Equatorial Guinea (Fig. 6.2). They found that the distinction between bushmeat, domestic meat and fish was much less significant than that between fresh and frozen produce. Frozen fish and chicken was cheap and widely eaten despite not being liked by consumers. Preferred foods included fresh fish and bushmeat. These were disproportionately eaten by wealthier consumers. As Bata's population and income increases due to its oil boom, it might be expected that consumption of these preferred foods would increase, with potentially severe consequences for sustainability.

Institutional framework—forest ownership in the Indian Himalayas

A case study from the Himalayas of Western India serves to illustrate the complexity of ownership and use rights that fall under the general heading of common-property resources (Berkes *et al.* 1998). The villagers recognised three kinds of private property agricultural land, four kinds of common-property grazing land and three kinds of forest land. The use of these land types by villagers varied with caste, gender and ethnicity. Use rights had been defined in 1878 under the colonial regime, based on traditional practices. However, new local resource management

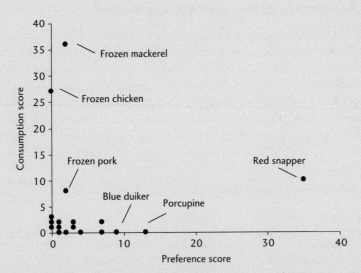

Fig. 6.2 The mismatch between consumption of foodstuffs and preferences for them, as expressed by consumers in the city of Bata. Fresh fish and bushmeat are much preferred, but frozen foods are overwhelmingly consumed, due to their low price and availability.

institutions are evolving; for example, the local women's organisation has been active since the 1980s resolving disputes and developing resource use rules.

Additional threats—threats to the anoa

The anoa, a dwarf buffalo endemic to the island of Sualwesi, is little-known, such that it is even unclear whether it is one or two species. Its range area has declined markedly compared with historical records, and this is driven by the synergistic effects of hunting for meat and loss of Sulawesi's forests. Between 1985 and 1997, 89% of Sulawesi's lowland forest was lost, leaving small fragmented anoa populations in the remaining forested areas. In those areas where habitat is not being destroyed, unsustainable hunting is instead widespread. The anoa has very few safe havens that are remote enough for neither of these threats to be operating (Burton *et al.* 2005).

In the next two chapters, we show how to use this information to put in place effective frameworks for sustainable use. In this chapter we discuss management tools that are commonly used to address the causes and consequences of over-use, while in Chapter 7, we look at how to implement management plans, monitor their conservation effectiveness and value for money, and ensure that they are resilient for the long term.

6.2 A taxonomy of management approaches

There is a bewildering array of methods for classifying types of management approach, not all of which are applicable in any particular situation, and several of which are slightly different perspectives on the same underlying issue (Table 6.1). It can be helpful to go through this typology though, as a way of clarifying the way that a potential management strategy is likely to influence hunting behaviour.

'**Carrot or stick**' addresses whether the manager wishes to modify the target's behaviour by imposing penalties for doing things that are frowned upon, or giving rewards for positive behaviour towards the resource. **Distractions**, on the other

Table 6.1 Different ways to classify management approaches.

Dimension	Options
Type of incentive	Carrot/Stick/Distraction
Point in the supply chain	Harvester/Trader/Consumer
Scale	Local/National/International
Target	Individuals/Communities/Institutions/Businesses
Focus	Top-down/Bottom-up
Implementation	Regulation/Persuasion
Addresses	Cause/Symptom

hand, aim to shift people's focus away from resource harvesting towards other, less damaging activities. **Top-down** interventions usually aim to change the behaviour of overarching institutions at the large scale (for example, by influencing government policy) and are usually initiatives from outside. **Bottom-up** interventions, on the other hand, are in theory ideas that come from the local people themselves, and can then be facilitated by outsiders. **Regulation/Persuasion** captures whether the intervention aims to change rules or to encourage people to comply voluntarily, while **cause/symptom** concerns whether the intervention is addressing the underlying issues causing unsustainable resource use, or whether it is focussed instead on controlling the resource use without fixing the underlying issues. Long-term sustainability requires that the causes of unsustainable use are addressed, although in the short run addressing the symptoms may be the only way to ensure species are not extirpated.

One management intervention might be supporting a local hunting cooperative in a village to set and manage sustainable hunting quotas (Box 6.3). The typology would then be as in Table 6.2. Another example is the listing of big-leaf mahogany on Appendix 2 of CITES, which requires that national management authorities only import or export the species after issuing a licence (Table 6.2; Blundell 2004). In order to issue the licence they need to make a 'non-detriment' finding that the sale of this shipment will not have a detrimental effect on the survival of the species in the wild.

Bushmeat hunting is one of the more intractable problems for policy-makers because it is multi-species, widely practised and informal, thus being extremely hard to govern (Milner-Gulland *et al.* 2003). Table 6.3 gives some examples of the types of policies that have been suggested to tackle unsustainable bushmeat hunting, with their pros and cons. These interventions address a range of components of the typology, and the table demonstrates how this way of thinking can lead to a more focussed assessment of management options.

6.3 How can we intervene?

Once we decide to intervene, there are a range of potential management approaches (Figure 6.3). In this section we look in more detail at the types of management intervention that target the people living with resources, and take a broad look at their underlying philosophy, strengths and weaknesses.

Assuming that conservation intervention only occurs when there is some current or potential issue of concern, people's behaviour has to be altered. However, there are two ways of doing this—either directly by controlling their resource use, or indirectly by changing their attitudes and opportunities so that they themselves change their use. If we intervene directly to control people's use of a resource, this must involve the setting and enforcing of rules. So we start by considering the issues surrounding rule-setting. Rules can either regulate use to a sustainable level, or prohibit use entirely. If use can be made financially, socially and ecologically

Table 6.2 Examples of the management typology in Table 6.1, for a project that is promoting community-based management of hunting in order to improve sustainability and for the listing of big-leaf mahogany on Appendix 2 of CITES in order to limit the international trade.

Dimension	Community hunting	Mahogany
Type of incentive	*Carrot and Stick.* Not a distraction as linked directly to the resource. Stick as includes rules, but external investment is the carrot.	*Stick.* This is a rule imposition, with penalties for illegal trade.
Point in the supply chain	*Harvester*	*Trader.* It targets the people trading timber between countries.
Scale	*Local*	*National/International.* National governments must implement the CITES legislation, though only for international rather than internal trade.
Target	*Individuals/Communities/Institutions.* Individuals are encouraged to set up a new community-based institution.	*Businesses.* Government institutions implement it but businesses are generally most affected.
Focus	*Bottom-up and top-down.* Government needs to be involved because of need to change land ownership laws. Degree to which it is bottom-up depends on how participatory the decision to go ahead with the management strategy is, and how involved local people are in its development.	*Top-down.* Based on a vote at CITES Conferences of the Parties.
Implementation	*Regulation.* Even though locally implemented, this is still based on setting and following rules.	*Regulation*
Addresses	*Symptom/Cause?* Depends on whether the lack of a properly regulated hunt was the key issue that needed addressing.	*Symptom?* There is no necessity for any conservation action to take place in-country, though the idea is that regulating the international trade will reduce unsustainable offtake.

Table 6.3 A summary of some of the policies that have been suggested for increasing the sustainability of bushmeat hunting.

Policy	Target	Effect	Pros	Cons	Reference
Reduce access to logging concessions by hunters (use of company vehicles, roads)	Logging company/ hunters	Increase hunter costs (of transportation to market and to hunting site).	Relative ease of implementation (small number of companies involved), may have large scale impact.	Top-down approach may lack grassroots support, relatively short-term.	Elkan and Elkan (2002)
Reduce demand for bushmeat by company employees (e.g. provide food)	Logging company/ consumers	Reduce hunter revenues (demand decreases, so price does)	Mitigates impacts of immigration. Easy to implement; may have impact on large scale.	Doesn't address general wealth and/or population increases in the area due to company presence.	Auzel and Wilkie (2000)
Education and awareness	All	Change attitudes.	Can change situation permanently without need for continual management. Relatively cheap.	Not always effective in short term. Individual responses differ. Ethical issues—cultural change and imposition of other values.	BCTF (2005)
Market inspections and arrests for illegal meats	Stallholders	Increase stallholder costs.	Can be very effective, particularly if few markets. Message clearly sent on which meats are acceptable to sell, which not.	Trade may shift location. Regular inspections needed on continuing basis.	Clayton et al. (2000)
Promote alternative protein sources (e.g. smallstock rearing, domestic meat)	Consumers/ Hunters	Increase opportunity costs of hunting; provide substitute good.	May be sustainable in long term, if alternative continues to be better option than hunting.	Link between conservation and livelihoods removed.	Brashares et al. (2004)
Checkpoints on roads (confiscation, arrest or taxation)	Hunters/ dealers	Increase hunter and dealer costs	Very effective if cannot be circumvented. Can selectively protect endangered species.	Must be continued regularly and indefinitely.	Lee et al. (2005)
Limit weapon types that can be used	Hunters	Decrease hunting efficiency, increasing	Protects traditional way of life. Can selectively protect	Not long-term solution; clock can't be turned back. Imposing	McClanahan et al. (2005)

Table 6.3 (Con't.)

Policy	Target	Effect	Pros	Cons	Reference
		hunter costs. Exclude outsiders.	endangered species.	unnecessary inefficiencies is not best approach to the problem except for protecting endangered species.	Whitman et al. (2004)
Restrict hunting effort in other way (e.g. closed seasons, size limits, protected species)	Hunters	Decrease offtake of vulnerable individuals.	Should have direct effect on wildlife population sizes. Can distinguish between resilient and vulnerable prey.	Needs monitoring and enforcement, may not be enough in itself.	
Offtake quotas	Hunters	Decrease offtake overall.	Direct link to biological sustainability. Can distinguish between species.	Needs high level of management.	Thorbjarnarson and Velasco (1999)
Ownership allocation (to individuals or communities)	Hunters	Change property rights.	Long-term solution.	Over-exploitation can still occur. May be difficult to legislate and conflict with existing rights.	van der Wal and Djoh (2001)
Patrolling protected areas (arrests, snare removal)	Hunters	Increase hunter costs.	Clear statement. Can protect populations directly.	May be resented by local community. Must be continued regularly and indefinitely.	Jachmann and Billiouw (1997)
Increase bushmeat production (ranching, domestication, food supplements)	Hunters	Increase potential offtake rates.	Improves livelihoods. Direct link can remain between production and forest conservation. Can promote community-level conservation.	May distort ecosystem dynamics for other species. May just increase profitability of hunting, so increasing hunter numbers. Domestication unlikely to work.	Feer (1991), Solis Rivera and Edwards (1998)
Designate no-take and extraction areas to improve hunting sustainability at landscape level	Hunters	Effects on costs and offtake unclear. Protects population.	Buffers against uncertainty and error. Clear designation of zones aids enforcement.	Still needs enforcement. Resentment likely if imposed. Yield improvements not guaranteed.	Beger et al. (2004), Hilborn et al. (2004)

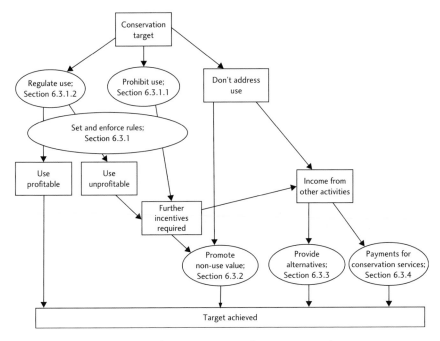

Fig. 6.3 Alternative approaches to managing the interactions between a conservation target and a community of resource users.

sustainable, then it can stand alone as a conservation solution. However, this is rare. If, as is more common, sustainable use is not possible, or not financially viable, then other strategies need to be employed. These are the same strategies as those available to address over-exploitation indirectly. We consider three of the main ones; enhancing cultural value and goodwill so that people are prepared to make sacrifices to conserve the conservation target; providing alternative livelihood opportunities that are not directly linked to the resource itself; and providing direct payments for conservation services.

In practice, this is not an either-or situation; several approaches are likely to be pursued in concert. The ones we consider here are also not the only kinds of management intervention, but they make interesting contrasts because their underlying philosophies are quite different. We focus on management interventions targeting people living locally with wildlife, rather than those aimed at people higher up the commodity chain (traders, end users) because local people interact most directly with their environment. Hence they will always need be part of sustainable conservation, even if others are also targeted for intervention. However, many of the principles outlined here are also applicable to other targets—for example, finding alternative livelihoods could just as well apply to traders as to hunters.

6.3.1 Setting and enforcing rules

Society imposes rules about most aspects of life, and biodiversity conservation is no different. Members of society are expected to comply with rules imposed for the benefit of society, and to face sanctions if they do not. The issue of whether conservation rules are fair and equitable is important, but the underlying assumption that social goods like biodiversity should have rules of use associated with them is one that is widely accepted. However, there is then the issue of who has the moral and the actual right to impose rules and sanctions on resource users. Non-governmental conservation organisations do not have these rights.

Enforcement of regulations is usually associated with top-down management by government, and has a heavy-handed and unfashionable reputation. It is often associated with prohibition of any kind of use. However, this is a narrow point of view that fails to recognise the range of types of regulation that can be used in conservation. Even a highly participatory community-based wildlife management association still needs to have rules by which members must abide, and sanctions for those who violate these rules. It is also not necessarily the case that individual resource users are the main target of regulations. Companies that export illegal timber on a large scale also need to face effective and well-designed sanctions, not just small-scale fuelwood collectors in National Parks.

The main key to the success of regulations is their **social acceptability**. Only if society at large buys into the necessity of controlling undesirable activities will it be possible to institute rules that are actually applied, and accepted, in practice. The threat of social opprobrium and ostracism can be a strong force discouraging people from breaking the rules. This can be seen in the case study of fuelwood gatherers in Lake Malawi National Park, discussed in Section 3.3.1.2—because the prevailing view was that fuelwood gathering was the local women's right, people did not buy permits to gather, and the Park Authority was unable to enforce the rule that all gatherers should hold one (Abbott and Mace 1999). In the UK, it is considered unexceptional to break speed limits but it is socially unacceptable to drink and drive, despite both being major causes of road mortality. This shows the power of public opinion—a campaign against drink-driving in the 1980s was effective in shifting society's norms.

The second key to success is ensuring that each actor is facing the **correct incentives**. This includes the potential violators of the rules, those who monitor them, and those who impose sanctions. There is a barrier to breaking rules imposed by social opprobrium, but once past this, we would expect people to weigh up the costs and benefits of their actions, and behave accordingly. Conservation projects often provide incentives to abide by the rules at the community level, for example, by providing clinics or schools. These benefits can help to make the community overall more positive about conservation, but individual hunters and monitors also need to see that the benefits to them personally outweigh the costs of working within the rules. So a monitor who does not see continued benefits from monitoring, even when there is no illegal activity going on, will stop monitoring, opening the way for illicit resource use to start again (Mesterton-Gibbons and Milner-Gulland 1998).

Analysis of the best way to structure regulations to ensure compliance uses tools from game theory (Section 5.4.2), cost-benefit analysis (Section 3.2.5.1), and the psychology of how people perceive risk. Game theory is important because people are acting strategically, and the best option depends on how other people behave. Managers need to try to avoid allowing free riders to get away with their activities. These are people who obtain the community-level benefits of a conservation programme along with everyone else, but do not bear the costs of cutting back on their resource use. This means that others bear the cost of the free-rider's activities. Risk perception is important because people's psychology varies depending on their personality and circumstances, and this affects their behaviour.

6.3.1.1 Prohibition of use

This means addressing over-exploitation by instituting and then enforcing regulations to prohibit people from carrying out environmentally damaging activities. The *steps* are:

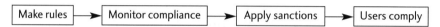

The *underlying philosophy* is that biodiversity is a social good, so society should decide on its use. Members of society should comply with rules imposed for the social good, and face sanctions if they do not.

Scope of the approach. Prohibition of use is just one extreme of a continuum of restrictions, going from no use, through limitations on use, to free access to resources for all. Despite its unsavoury reputation, prohibition of use is probably still the most widely used and identified-with approach in conservation. Protected areas with regulations governing public access, banning international trade in endangered species, bans on the use of particular gear types or on commercial sale of particular species are all ubiquitous management tools.

Keys to success. Because prohibition of use is an extreme measure, it is likely to require a particularly high level of **social acceptability** in order to ensure that compliance is high. It is also important that people perceive that there will be **serious consequences** if they break the rules. Law enforcement can become ineffective if people start to realise that they will not actually face sanctions; this is what happened with market patrols in North Sulawesi, when the reduction in illegal babirusa sales was less marked with each patrol because people were not being arrested (Box 3.2).

The *framework for analysis* of prohibition is the same as for setting and enforcing rules more generally (covered in Section 6.3.1). Cost-benefit analysis allows us to weigh up the costs and benefits that individuals face, and predict whether resource use is worthwhile or not. For example, in the Luangwa Valley in the 1980s, the benefits from elephant and rhino hunting far outweighed the costs for commercial poachers with good weapons, but hunting was not worthwhile for local people who could not reliably kill an elephant with their muzzle-loading guns (Leader-Williams and Milner-Gulland 1993; Box 6.2).

Box 6.2 Effective law enforcement in the Luangwa Valley, Zambia.

Jachmann and Billiouw (1997) investigated the effect of law enforcement budgets, and their allocation, on the number of elephants found killed illegally in the period 1988–1995. They found that illegal elephant kills were very well predicted by a model including the number of bonuses paid to scouts and the number of effective patrol days. The more patrol days and the more bonuses were paid, the fewer elephants were found killed. They had good reason to believe that all animals killed were found, so there was no effect of patrol effort on detection of poaching, suggesting that the relationship between law enforcement and deterrence of poaching was genuinely being measured by carcass counts. They also argued that the number of bonuses paid was related to the size of the bonus on offer, such that scouts tried harder when bonus rates were higher. The key messages are that rewarding individuals for information and arrests is highly effective; that investigations outside the park following up leads are about four times more cost-effective than foot patrols within the park (Figure 6.4); and that the overall law enforcement budget is an important determinant of illegal activity.

Fig. 6.4 Information-gathering with hunters, Congo. Photo © Pat Aust

Issues with this approach to management include how we define 'society', and hence who has the right to impose potentially highly restrictive rules on resource users. For example, does society include the interests of those in rich nations who wish to see their moral standards applied throughout the world? The issue of whaling exemplifies the different world-views that exist concerning the acceptability of use of nature. In 1992, a poll coordinated by Gallup Canada asked people's opinions about the statement that 'there is nothing wrong with whaling if it is properly regulated'. About two-thirds of respondents in the UK and Australia disagreed with this statement, while three-quarters of Norwegians and two-thirds of Japanese respondents agreed with it (Freeman 1994).

In general, prohibition of use addresses the **symptom** of over-exploitation, but not the cause. Unless the cause (for example, poverty, international demand, cultural values) is addressed, this approach is just sticking-plaster. That is why we need to give people further incentives to conserve, rather than relying solely on prohibition.

6.3.1.2 Regulated resource use

This involves bringing the use of the resource within sustainable limits. The *steps* are:

The *underlying philosophy* is that it is in the interests of both society and users for resources to be maintained through sustainable use. This is a materialistic perspective, whereby the resource is valued for its products rather than its intrinsic value.

The *scope* of the approach is broad. It is the fundamental basis of fisheries and wildlife management. The usual strategy is to take the principles from these commercially based resource use operations, and apply them to conservation situations where Western regulatory institutions have not previously been used. Direct regulation of resource use can occur at all levels: CITES aims to regulate resource use at the international level, by agreeing export quotas for species that might be threatened by trade. At the other extreme are village-level hunting cooperatives which are formed with the aim of monitoring and controlling use of their resources, whether it be for trophy hunting or subsistence (Box 6.3). There are many regulatory tools available which have been tried in other sectors. Use can also be non-consumptive, for example, ecotourism based on the resource's aesthetic value.

The *keys to success* are:

- The fact that the object of conservation concern is the focus of the management intervention means that conservation and livelihood benefits are explicitly linked. This focus is an advantage of this approach, so long as the resource is able to support the **level of harvest** that is required to satisfy livelihood needs. If the resource is overexploited at the outset, which is almost invariably the case, then in the short term there will need to be a period of reduced offtake while the resource recovers. There will need to be some livelihood subsidy to take users through this period. There is an expectation that by the end of this recovery period, the new regulated offtake levels will be large enough to sustain the costs of management and still provide adequate returns to users. This requirement needs to be explicitly addressed, rather than assuming that once the resource is being sustainably harvested, all will be well.

- Even without explicit effort control, **legalisation** of resource use, bringing an informal trade into the formal economy, can be useful if it leads to recognition of the true value of the resource by government and users. Once a trade is in the formal economy it can be taxed and regulated. However, the legal use may mask continuing unregulated use, and so unless enforcement is effective, legalisation does not take management any further forward.

- Correct **ownership and control** of the resource are vital. Users must feel security of tenure, to ensure that they buy into the need for sustainable use for the future. Managers must have the power to act if use appears unsustainable.

- The resource and the institutional framework need quite **specific characteristics** if regulated resource use is to work. For example, particularly suitable resources are resilient so they can withstand high levels of use, have a relatively high value/volume ratio, are non-perishable and have an accessible market (Salafsky *et al.* 1993; Box 4.6).

The *framework for analysis* depends on whether you are looking from the manager's or the user's perspective. From the manager's perspective, the issue is what the social optimum is (i.e. what is best for society), and how best we should get there. Bio-economics provides the foundation for this analysis; see Cochrane (2002). However, in conservation, the perspective of the natural resource manager is not always entirely appropriate. Rather than being a commercial industry, many uses that are of conservation concern are informal and culturally embedded. Hence some of the other frameworks listed in Table 3.1, which are focussed on the user, are also useful.

One of the main *issues* is that conserving biodiversity through sanctioning the direct use of the endangered resource can be controversial and potentially risky. There is a risk to the species if the management fails to reduce use to a sustainable level, and a public relations risk to the implementer, regardless of the potential benefits of the scheme. Every time that the parties to CITES meet, there is a row about the ivory trade, which has little to do with the sustainability of use, and much more to do with opposing ethical perspectives on whether it is acceptable to kill elephants. This is clearly demonstrated in the press releases put out by the Born Free Foundation and the International Council for Game and Wildlife Conservation when the ivory issue was discussed at the 2002 CITES meeting, entitled respectively 'Stop the slaughter: ban the bloody ivory trade' and 'Sustainable hunting: an instrument for species protection and to fight poverty' (Born Free Foundation 2002; CIC 2002). IUCN—the World Conservation Union—has adopted a policy statement on the sustainable use of wildlife, which has also attracted substantial debate (IUCN 2000).

Another concern is that by taking a use-based perspective, we may lose sight of the importance of other societal values for conservation. For example, Oates

Box 6.3 A community-based hunting association in Cameroon.

The Mokoko Wildlife Management Association (MWMA) was formed in 1997 by a group of communities, with facilitation from the Mount Cameroon Project (http://www.mcbcclimbe.org/mcp_intro.shtml). The Association was legally recognised by the Government of Cameroon as authorised to manage wildlife within their designated area, and to hunt for personal consumption. This included making detailed rules for MWMA members, and fining violators. They developed a scheme such that hunters were allocated quotas in the form of metal

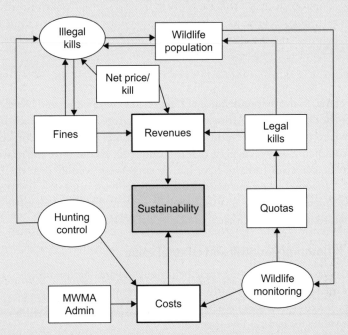

Fig. 6.5 A simplified representation of the financial position of the MWMA. Improved revenues can come from an increase in the number of bushmeat kills, the price/kill net of hunting costs (assuming MWMA takes a cut from the sale price of each animal killed rather than charging hunters per tag), and the number and level of fines. The number of fines and the number of animals killed legally and illegally depend partly on the effort expended in hunting control, which is the main variable cost to MWMA. The driving external factors, which are influenced but not fully determined by hunting control, are the size of the wildlife population and illegal hunting effort. Ovals represent key decisions: the amount invested in hunting control vs monitoring by the Association, and the amount of illegal hunting by poachers; rectangles represent the consequence of decisions or factors out of the Association's control. Adapted from Olsen *et al.* (2001).

tags which had to be attached to each animal killed. The quotas were agreed following participatory animal abundance estimation by hunters, complemented by transect surveys by external scientists (which produced very similar results).

MWMA set up monitoring of both hunters and the resource, hoping to turn a depleted and unmanaged wildlife population into a viable concern. A cost-benefit analysis suggested that the Association would only be self-sustaining in the long term if the monitoring system was given start-up funding from outside, and resulted in a large proportion of violators being discovered and managers receiving at least 50% of the resulting fines. Long-term viability also depended on wildlife populations recovering and yielding increased offtakes to hunters. Hence although the MWMA was very successful in terms of buy-in by local people, its actual sustainability in the face of management costs was unclear (Olsen *et al.* 2001; Figure 6.5).

(1999) contrasts the preponderance of Integrated Conservation and Development Projects (ICDPs) in West Africa with the more protectionist approach taken in India, where religious and cultural reasons for conservation are accorded much more prominence. Oates suggests that this difference accounts for the relative success that he perceives India to have had in conserving wildlife over the last few decades. However, in recent years, India's protectionist approach has started to show weakness, partly due to the inequities perceived by local people excluded from their resources (e.g. Chhatre and Saberwal 2005).

6.3.2 Promoting goodwill and cultural value

This involves people changing their behaviour towards nature because they feel that it is the right thing to do, rather than for economic reasons. They are prepared to accept the costs that they incur from this change in behaviour because they hold non-use values for nature. The *steps* are:

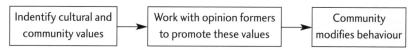

The *underlying philosophy* is that the most powerful reasons for changing our behaviour towards nature are not derived from material benefits. People respect their environment for its spiritual significance or to abide by social norms. This respect can be built through education, which can take many forms—from school-based learning through public lectures, exhibitions, film shows and displays, to working to change the views of opinion-formers in society so that they spread the word that certain behaviour will not be tolerated. Different societies have different conventions about the relationship between humans and wildlife, both in terms of individual animal welfare and the environment more broadly; hence different approaches are needed. It would be difficult, for example, to convince a trader in

Sulawesi that carrying live domestic dogs to the market for consumption is a breach of animal welfare standards—in the UK, by contrast, the dogs in Figure 6.6 had to be air-brushed out when the picture was used in a fund-raising campaign, for fear that donors would be upset. The wild pigs carried alongside the dogs are only eaten in the Christian tip of the island because the rest of the island, being Muslim, does not touch or consume pigs, which are seen as unclean—this religious taboo has acted like a refuge for the wild pigs, meaning that they have been hunted much less intensely in the past than might have been expected (Clayton *et al.* 1997).

The *scope* of the approach encompasses two main types of interventions. The first is highlighting the **non-use values** of nature. Under this heading are all the public awareness and dissemination activities that highlight the beauty of the natural world and our responsibility towards it. Another set of conservation projects has a different approach—they build **community goodwill** towards the conservation project or its organisation and personnel, rather than directly towards the object of their activities. In this way the community feels predisposed towards helping the organisation fulfil its aims. This is not the same as providing alternatives, because the conservation organisation is explicitly not aiming to compensate for economic value lost from refraining from harmful actions. Instead the organisation demonstrates its commitment to the community, and to helping them with their needs. In the longer term, the community might also then start to

Fig. 6.6 A wild pig dealer with his truck at a checkpoint in North Sulawesi. On the ground are pieces of several endemic species (babirusa *Babyrousa babyrussa*, Sulawesi Wild Pig *Sus celebensis* and dwarf anoa *Bubalus depressicornis*), on the truck are domestic dogs also destined for market. Photo © Lynn Clayton.

hold its own intrinsic values for the conservation object, through recognising that others see it as important. An example of a successful implementation of this approach is in Box 6.4.

The *keys to success* are:

- Making interventions **culturally appropriate** and relevant to the communities involved. This means having a deep understanding of their needs and

---

Box 6.4 Building goodwill in Pipar, Nepal.

The World Pheasant Association has worked with villagers in the mountainous region of Annapurna for over 25 years. Based on discussions with local people, they built and have maintained schools in the valley in return for villagers protecting local forests, home to five Himalayan pheasant species (Figure 6.7). In 25 years, the populations of these pheasants have remained stable in the WPA area, while declining elsewhere in the region, suggesting that people are holding to the agreement. A recent review suggested that villagers who had been involved with WPA for longest were clear about the conservation basis for WPA's involvement in the area, but more recently contacted villages had not perceived the link and some thought that WPA was a development organisation, suggesting that more work was needed to reinforce the link between pheasant conservation and community benefits. The WPA's general approach is to emphasise the cultural significance of pheasants, and to build goodwill rather than focus too strongly on use-based projects. This has also paid dividends in the Palas Valley, Pakistan, where the WPA was able to respond quickly to people's needs after the Asian earthquake in 2005, due to its long-running presence in the area, earning the organisation the trust and respect of local people.

Source: www.pheasant.org.uk.

(a)   (b)

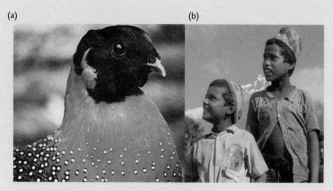

Fig. 6.7 a) Satyr tragopan Photo © Jean Howman/WPA, b) School children Photo © Dick Potts/WPA.

aspirations, and of how they relate to nature. Imposing Western values will not work.

- Ensuring that the key local **opinion-formers** are engaged and are convinced of the value of the approach.
- Making sure that interventions promoting goodwill are providing services that the community wants, and particularly that those whose behaviour you are trying to influence **benefit** from the interventions (for example, if hunters don't benefit from a grain storage facility, providing it will improve relations with the village but may not change the incentives of the people who matter most for conservation success).

The *framework for analysis* is predominately sociological and psychological, understanding how people's attitudes and opinions are shaped, and how these affect their behaviour (e.g. Tanner 1999; Bamberg 2003). In a study of factors affecting environmental behaviour towards water in the Scottish Highlands, Spash *et al.* (in press) show that people's behaviour is most strongly determined by their ethical beliefs and their perceptions of the social dynamics of changing behaviour, rather than by economic costs and benefits. Demonstrating the link between **attitudes and behaviour** is often difficult in conservation, where the activities of interest may be illegal, and changes in the status of exploited species are hard to assign to changes in an individual person's behaviour. Unsurprisingly, therefore, when Holmes (2003) reviewed the literature on local communities' relationships to protected areas he found that, although there are a number of studies looking at attitudes to conservation, very few carry this through to demonstrate that attitudinal change has actually led to behavioural change, and hence to conservation success.

There are two main *issues* with this approach. Although promoting community goodwill and the non-use value of nature through education and public awareness-raising are ubiquitous components of conservation, there has been no proper **analysis** of how best to approach promoting these values, or of the cost-effectiveness of different interventions. It may be that leaflets, films and talks interest people, and may change their short-term attitudes, but cannot in themselves affect behaviour—in which case are they worth substantial investments of time and money? Because virtually all conservation projects have these components, and they are not usually clearly costed and monitored, it is very difficult to disentangle their effects from all the project's other activities. Examples such as the World Pheasant Association's intervention discussed in Box 6.4 are important and particularly useful because there are fewer confounding variables than usual.

The second issue is that goodwill gestures and raising concern for the environment are mainly likely to be useful in situations where the **costs** of refraining from a damaging activity are low. They are unlikely to work in isolation when the activity is a substantial contributor to livelihoods. In these cases, economic necessity may override cultural value. However, even this has been shown not to be entirely true when non-use values are very high. Even in extreme conflict situations local people have continued to protect National Parks, at great risk to their lives. For example, during the Rwandan genocide, the Virunga National Park was protected by local rangers after the evacuation of expatriate staff, ensuring the safety of the mountain gorilla population (Hart and Hart 2003).

6.3.3 Alternative livelihoods

This means enabling people to stop over-using natural resources by giving them another way to make their living (also called distractions—see the typology in Table 6.1). The steps involved are:

The *underlying philosophy* is that individuals have the moral right to choose to make their livelihoods as they wish, within local societal norms. If wider society wishes them to change their behaviour, then a viable alternative activity must be made available to them. The move away from natural resource use should be **voluntary and costless** (ideally beneficial) to the users.

The *scope* of this approach is most usually small-scale conservation at the village level. Often external NGOs initiate the management activity. Partly this is because they have no power to institute and enforce regulations, and so offering alternative livelihoods and public awareness are the only ways in which they can realistically intervene. The approach is very popular nowadays, because it avoids some of the moral issues that are inherent in imposing and enforcing regulations on people who are bearing the cost of living with wildlife for the benefit of a wider (usually Western) society. By offering alternative livelihoods, NGOs hope to ensure that conservation has **popular support**, and is seen as going hand-in-hand with development (Box 6.5). Integrated Conservation and Development Projects (ICDPs or ICADs) are seen as the ethical, effective and people-orientated way forward for conservation, and often have the provision of alternatives at their heart. There has been criticism of the ICDP approach from both the development (e.g. Sekhran 1996) and conservation sides (Oates 1999), and they are certainly not an easy option. As an example of the issues that ICDPs must struggle to overcome, Vrije Universiteit's (2001) evaluation of the Mount Elgon ICDP highlighted administrative problems which delayed the receipt of equipment by project officers on the ground; local people's attitudes of dependency due to previous development initiatives in the area; progressive reduction of commitment to the project by external donors; inflexible planning; and flawed project design. Although ICDPs are usually associated with small-scale conservation in areas where local people's activities are perceived to be the main threat to biodiversity, distraction in the broader sense could also include finding alternative products for natural resources such as timber or plant oils, or consumers switching away from traditional medicines towards synthetic alternatives.

Box 6.5 Integrated Conservation and Development for Bwindi impenetrable forest

The Bwindi Impenetrable National Park is home to one of the few populations of the endangered mountain gorilla *Gorilla berengei*. By the 1980s, the area's gorilla population was threatened by large numbers of people entering the forest for timber extraction, snaring and mining. An integrated conservation and development project was run at Bwindi in 1987–2002 by CARE-Uganda, funded by USAID and focussed mainly on improving agricultural productivity. The Park was gazetted in 1991, causing resentment from local people because of their exclusion from valuable forest resources. This conflict led to the establishment of zones in the Park where limited resource harvesting was allowed, including bee-keeping and gathering of herbal medicines and basket-making materials. This and other initiatives led to better relations between Park authorities and local people, particularly the bee-keepers (Fig. 6.8). However, the focus on community-based natural resource use programmes at the edge of the Park, aimed at bolstering public support for the Park and its role in protecting gorillas, meant that less time was spent patrolling the core area of the Park. This is where the gorillas are mostly found, because they avoid human disturbance. Here snaring rates continued to be high, threatening the long-term sustainability of the gorilla population. So although the ICDP was successful from the perspective of improving local support for conservation, gorilla conservation did not benefit as much as it could have done, because human disturbance and snaring remained a

Fig. 6.8 Beekeepers at Bwindi have benefitted from being allowed into the Park's limited use zones. Photo © Julia Baker.

problem, and ranger attention was diverted away from the core areas of the Park. This study demonstrates how important it is not to lose sight of biodiversity-based indicators of conservation success, particularly those associated with the primary goal of the intervention (protecting gorillas, in this case), as well as monitoring social indicators.

Source: Baker (2004).

The *keys to success* are:

- Effective **adaptive management** and learning from mistakes are vital in projects that aim to change people's behaviour towards natural resources voluntarily, through offering alternatives. This is true for all types of management, but often most acute in this indirect conservation approach. We will discuss adaptive management at length in Chapter 7.
- There needs to be an **acceptable and viable alternative** available that is financially, socially and environmentally sustainable and resilient to shocks. International mass tourism, for example, is vulnerable to civil unrest, while there may be few alternative options to natural resources as a last resort support to people's livelihoods in difficult times.
- The **economics** need to be right. There needs to be a fixed supply of labour, such that when labour is switched from the unsustainable natural resource use to another activity, resource use actually declines. In addition, switching to the alternative should not drive up demand for the resource or enable more efficient harvesting. Damania *et al.* (2005) use a model to show that investment in improving agricultural production can potentially increase bushmeat hunting rates. This happens both because as people's incomes improve, they demand more meat, so driving up revenues from bushmeat hunting, and because they are then able to invest in more efficient hunting technology, such as guns. Auzel and Wilkie (2000) demonstrate this effect empirically; following the arrival of a logging company in an area, 49% of local villagers' meals contained bushmeat, compared with 39% in villages further away. This was because of the extra income that they gained from servicing the logging camp, allowing them to spend more on food. Most of their income came from supplying workers with bushmeat; because they had disposable incomes, logging workers' meals contained bushmeat 76% of the time. Oates (1995) argues that an ICDP in the buffer zone of Okomu Forest Reserve in Nigeria improved the standard of living there, leading to immigration to the area and to further destruction of the biodiversity that the project was put in place to conserve. Sievanen *et al.* (2005) show that the introduction of seaweed farming had very mixed effects on fishing effort, in part because it did not fully substitute for fishing as a livelihood activity.
- In theory, there doesn't need to be an **explicit linkage** between the alternative activity and conservation for it to work, at least in the short term, so long as the

presence of the alternative means that people spend less time harvesting the resource. But in the long term, a lack of linkage means that other threats may not be averted, or the original threat may resurface when the alternative becomes less attractive. For example, if villagers are diverted into smallstock raising, they may no longer have any interest in preserving forested areas for their bushmeat content, increasing the likelihood that these areas will be logged or converted for agriculture. This concern for long-term sustainability is why it is important to link alternative livelihoods explicitly to conservation within an ICDP.

- By definition the alternatives are initially less attractive than the unsustainable resource use, otherwise people would already be doing them. There is thus a need to identify the **reasons why** people are not already diversifying away from natural resources. If there is a simple barrier such as lack of access to microcredit, training or suitable markets then external intervention can act to lift it. It might be that a single capital injection would lead to sustained viability of alternatives (such as building ecotourist lodges and paying for the initial publicity). More usually, analysis suggests that alternatives need to be permanently subsidised in order to make them attractive to people. This then increases incomes, potentially causing more pressure on resources, and may lead to inwards migration to the area. Unless the underlying causes of over-dependence on a declining natural resource base are identified and addressed, alternative livelihoods will only ever be a stop-gap measure that fails once support is removed.

The *framework for analysis* works at two scales. Household economics enables us to analyse how individual households allocate their labour between competing activities (e.g. Barrett and Arcese 1998). Institutional analysis looks at how institutions can best be structured to ensure that the incentives for good management are correct (Anderies *et al.* 2004). The **sustainable livelihoods approach**, which is widely used in development, is a qualitative approach that identifies the aspects of people's livelihoods that are vulnerable to stresses (DFID 2001). It is useful as a framework for baseline studies, but has less to say about how best to structure interventions such as ICDPs.

The *issues* inherent in this approach are at the opposite extreme to those raised by the sticks approach of preventing use or setting rules. When is it appropriate to provide alternatives to people who are causing costs to wider society by damaging the public benefit of biodiversity? How do we define the user group which is entitled to these alternative benefits? For example, when does a settler or itinerant hunter become a 'local' person? The ethics of development intervention demand that no one should be disadvantaged by your actions (see Chapter 3), but the control of resource use for the benefit of wider society always has costs; the question then becomes who should bear these costs (Norton-Griffiths and Southey 1995). A good example of the context-dependence of ethical viewpoints about who is considered to have the right to use resources is **bushmeat hunting**. Hunting bushmeat for commercial sale is seen by some in the West as undesirable, while those who use it for subsistence are seen to have rights that need defending (for example, the Bushmeat

Crisis Task Force's goal is to 'eliminate the illegal *commercial* bushmeat trade...' [our italics]; BCTF, n.d.). In Sarawak, Malaysia, government action to ban the commercial sale of wild meat was welcomed by local people, who feared the loss of their subsistence livelihoods (Tisen *et al.* 1999). But evidence from the Democratic Republic of Congo shows that very poor people often rely on bushmeat sales as one of the few accessible sources of cash, needed to buy cheaper foodstuffs and for expenses such as school fees and medical care (de Merode *et al.* 2004). In what way, then, would it be more ethical to stop the use of bushmeat as a source of cash, rather than food?

6.3.4 Payment for conservation services

This involves paying people directly for their contribution to conservation. The *steps* are:

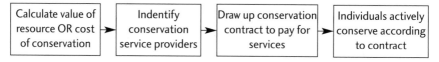

The *underlying philosophy* is that by conserving, or not using, natural resources, individuals are incurring costs. Active conservation has direct costs, and refraining from use has opportunity costs. However, the majority of the benefits from conservation are accruing to others, particularly to those who value nature and wish it to be conserved. There is a need to balance the costs and benefits of conservation, so that those of us who benefit from conservation pay its true costs, and those who are bearing the costs are compensated (Norton-Griffiths and Southey 1995; Balmford and Whitten 2003). Supporting alternative livelihoods can be seen as compensation for opportunity costs, and supporting regulated use of the resource is offsetting active management costs. However, some argue that these approaches are economically inefficient and difficult to implement successfully—a better approach would be to **directly compensate** people for the costs they incur in conserving their local resources (Ferraro 2001; Ferraro and Kiss 2002).

The *scope* of this approach is broad, and growing rapidly. It encompasses a range of mechanisms from competitions with cash prizes, through contracts with landowners to manage their land for conservation, to quota trading schemes. The unifying theme is that the value of the actions of individuals who can directly influence conservation outcomes are recognised and financially rewarded. The approach is analogous to the single farm payments introduced by the European Union in 2003, breaking the link between agricultural production and subsidies, in order that farm payments could be given specifically for farmers' costs in providing social goods such as biodiversity conservation (European Commission 2003). In Australia, **conservation auctions** are being piloted, whereby farmers bid for funding to carry out conservation on their land (Gole *et al.* 2005). The most famous developing country scheme is in Costa Rica, where government has entered into conservation contracts with landowners, in a Payments for Environmental Services Program designed to reduce forest clearance (Zbinden and Lee 2005).

The *keys to success* are:

- Most critically, there need to be people whose activities influence conservation success and who are in a position to sign contracts that commit them to constraints on their behaviour. This generally means that they must have **property rights** over the area of conservation interest. Hence in Costa Rica and Australia, farmers can sign contracts to manage their land a certain way, and these are easily monitored and enforced. However, in developing countries, property rights are not always clear and enforceable. This is not necessarily an insurmountable obstacle, however. Contracts can be signed with communities rather than individuals (Durbin 2003), or can include a commitment to change one's behaviour and influence others, rather than to guarantee a conservation outcome (Box 6.6).
- **Government backing** is required in order to ensure that conservation contracts are enforceable in law. This may be difficult to achieve if it involves granting rights to communities that were not previously present.
- Agreements should be fairly **negotiated** with local people, and contribute to building long-term sustainable livelihoods, rather than being seen as a cheap and quick way of obtaining conservation benefits (Romero and Andrade 2004).

The *framework for analysis* of direct payments for conservation is **market-based instruments** (MBIs). These have a long history in environmental economics, having been widely applied in pollution control. The idea there is that individuals or firms act in an environmentally damaging way because they are not paying the full costs of their actions—they generate **externalities** which are costs incurred by people other than the person carrying out the damaging action. Buying a permit to pollute transfers these costs back to the polluter (Field and Field 1997). Turning

---

Box 6.6 Conserving snow leopards (*Uncia uncia*) in Mongolia.

The International Snow Leopard Trust (ISLT) has used conservation contracts successfully. In 1998 they started Snow Leopard Enterprises in Mongolia, where they enter into a contract with herders to buy woollen handicrafts at fair trade prices (Figure 6.9). The contract is signed in the autumn, and herders agree not to poach either snow leopards or their prey. In the spring, when the products are purchased, they receive a 20% bonus if no poaching incidents have been recorded in the area. This provides an incentive not just to refrain from poaching themselves, but to stop others poaching. In 2003, the scheme covered 50% of the snow leopard's range in Mongolia, had expanded to Kyrgystan, and there were no reported poaching incidents. Although this may look superficially like an ICDP-type approach, the difference is that there is a contract to buy goods, rather than intervention to set up alternative livelihoods. The direct payment for conservation is not the purchase of the goods per se, but the bonus scheme.

Fig. 6.9 A pile of handicrafts. Photo © Terry Blumer. Courtesy of the Snow Leopard Trust

Sources: http://epp.gsu.edu/pferraro/special/SnowLeopardMongolia.pdf; http://www.snowleopard.org/about/resources/2003annualreport.

this idea on its head, we get **rights-based approaches** to fisheries management. These involve giving individuals or communities rights to fish, using **individual transferable quotas**, for example (Charles 2002). The number of quotas issued depends on the sustainable level of overall fishing, but because a market for quotas is created, the price of a unit of quota depends on its value—i.e. the value of the fish that can be caught using it.

Direct payments for conservation services fit within the MBI framework because a market is created for conservation services where none existed before—people's conservation actions are given monetary value based on what conservationists or governments are willing to pay. This allows externalities to be expressed monetarily; in this case the local people are bearing costs and the global community are obtaining benefits from conservation. Because at the moment it seems that conservation costs, although unfairly distributed, are far below the potential benefits (Balmford and Whitten 2003), the use of MBIs should increase conservation levels considerably.

A number of *issues* have been raised concerning the use of direct payments for conservation. In many cases, the difficulties involved in direct payments are similar to those encountered for ICDPs, but are more explicit. For example, it is obvious that direct payments require **long-term financing** in order to ensure that conservation continues to happen. However, the chances of ICDPs or sustainable use schemes breaking even are also not good in the majority of cases, so they too will need long-term external support. There are costs involved in setting up the infrastructure for developing and managing the contracts, but this too is likely to be similar to that required for other kinds of intervention. An analysis of the costs involved in the Australian conservation auction pilot scheme suggested that, although initial set-up costs were high, overall the scheme was 2–3 times more efficient than the alternative of paying a fixed price for a given conservation service regardless of the costs that farmers actually incurred (Gole *et al.* 2005). The proponents of direct payments state that they are flexible, efficient, explicit about the reasons behind the intervention, and so are simple to understand. Opponents are concerned for a number of reasons (Romero and Andrade 2004). They worry about **commercialising** nature by introducing market mechanisms when nature is so hard to value, and possibly weakening other, non-use values for nature. They are concerned about the **power asymmetry** that exists in the negotiations between local people and international conservation organisations. This means that local people may not get a fair price for their conservation services and may lose their rights, and that direct payments schemes may not help to build local **social capital** (such as community groups or education) and empower people to improve their livelihoods through engaging them in rural development. ICDPs and sustainable use schemes have the advantage that they explicitly aim to build local institutions, which will then have a knock-on effect on community empowerment more generally.

The issue about underpaying comes down to whether schemes pay people's **opportunity costs** for conservation, or pass on the benefits from conservation that outsiders receive. Opportunity costs can be relatively low in poor areas and comparatively easy to quantify, whereas the values of nature are likely to be both large and very hard to monetise. However, it seems likely that any conservation intervention will only pay what is needed to conserve wildlife, rather than investing at a level reflecting true global biodiversity value, so again the issue is whether direct payments are just making this inequity explicit. There is also no reason why direct payments can't improve social institutions and empower local people if they are well implemented (Durbin 2003).

6.3.5 Which approach?

The four approaches discussed here all have strengths and weaknesses, which make them more or less suitable to particular circumstances. Referring back to the data requirements discussed in Part 6.1, we can suggest when each approach is most appropriate.

Regulated use requires a resilient and productive species, providing a marketable product. Success is more likely if revenues can be generated quickly and monitoring is relatively cheap. There is a need for stable and well-developed institutional structures, such as a resource users' association, that can implement the management plan. Few species of conservation concern meet these requirements, although it may be possible to use an umbrella species and conserve the other species that share its habitat.

Alternatives require realistic robust livelihood activities that can truly replace the damaging activity (rather than just supplementing, or even stimulating, it).

They require a defined community so that outsiders are not attracted by the new opportunities. The approach is more resilient if there is some linkage with the conservation target, so that when the situation changes, the damaging activity doesn't resume. There is a need for institution-building to support the new activity. These conditions make finding feasible alternatives that fully compensate for the loss of livelihoods from resource use difficult, although alternatives can make a useful component of a conservation package (Ireland *et al.* 2004).

Direct payments need users to have the ability to control resource use on their land, whether they are a community or a private individual. Binding long-term contracts are needed for all parties. Power imbalances need to be corrected for, so that all concerned enter into contracts by free will. These payments should promote, rather than undermine, people's intrinsic value for nature. Direct payments are potentially a flexible and widely applicable mechanism, being particularly suitable for resources that have little use value for their owners but may have high value to others (such as watershed protection through conservation of forested slopes). This may mean, however, that they are less often applicable to exploited species.

Promoting goodwill is a critical part of all conservation activity. On its own it is likely to work best when the costs of conservation are not high, and when the issue is not contentious. These situations are not the norm, and are likely to continue to reduce in number as environmental degradation, development and population pressures build up.

6.4 Implementation strategies

Having decided which general approach to take, the next question is how best to implement the approach in practice. How can we translate a conservation philosophy into practice? We focus here on the human side, and don't attempt to address issues within conservation biology—such as how best to restore degraded ecosystems, how to manage small populations, how to ensure habitat connectivity, or the relative merits of species-based or ecosystem-based approaches to management. These are subjects for a different book (e.g. Sinclair *et al.* 2005; Primack 2006).

6.4.1 Direct use

There is a wide range of regulatory instruments available for controlling natural resource use; see Cochrane (2002) for an excellent review focussed on artisanal fisheries. Some of the commonest ones are:

- Restrictions on the amount harvested. This is the most direct way to implement sustainable hunting levels. At its most basic, a harvest restriction involves setting a **Total Allowable Catch** each year and allowing people to use the resource until this offtake level is reached. The TAC is based on calculations of the sustainable harvest level. The approach is not suitable for use alone because it promotes open access behaviour by harvesters. There is an incentive for each individual to harvest as much as possible of the resource as quickly as possible before others can do the same. For example, in the early 1990s, it took 24 h for the Alaskan halibut fleet to catch their annual TAC (Barlow and Bakke n.d.). However, it is generally necessary to have a TAC underpinning other management measures. More workable catch restrictions include giving individual harvesters the right to take a proportion of the TAC, or to kill a certain number of individuals each year. If these permits are then made tradeable (called an **Individual Transferable Quota**), a market can develop for them, which can give substantial advantages in terms of economic efficiency (Kerr *et al.* 2003; Job Monkey 2005).

- There are lots of restrictions possible on the type of individual caught. These can include the species caught (e.g. resilient species only), sex (perhaps males only), age (e.g. adults only) or size class (over a certain length or weight). One key problem with these kinds of restrictions is **by-catch**—people unintentionally catching the wrong type, and then having either to discard it or sell it illegally—discards being wasteful and still contributing to mortality of the protected component of the population, and illegal sales undermining the whole management plan. Even if by-catch is avoided, there is a need for quite a sophisticated understanding of population dynamics in order to ensure that these management rules have the desired effect on sustainability of resource use. For example, catching only males may work at low levels of harvest, depending on the mating system, but may cause reproductive collapse at higher harvest levels (Ginsberg and Milner-Gulland 1994). Restricting catches to larger individuals can be useful if it ensures enough individuals reach reproductive age, but in some species (such as fish), the largest individuals are also the most fecund, and removing them can have a disproportionately large effect on population growth. Using restrictions on the type of individual caught in conjunction with other measures can produce a more nuanced management strategy, however. For example, giving out permits to kill only a certain number of full-grown adult males per year is the usual management method for trophy hunting (Whitman *et al.* 2004).

- The majority of management plans involve **effort restrictions**, which limit the amount or type of effort that people can put into harvesting. This then indirectly limits the amount of catch they can obtain. Typical effort restrictions include granting licences to access the resource and limiting the type of gear that people can use. Effort restrictions are widely used in fisheries because they are relatively easy to enforce; it can be easier to monitor whether people have a permit to use a resource, and what type of weapon or boat they are using, than to monitor their offtake. However, there is a strong incentive to circumvent them. People will plough their profits from harvesting back into improving their technology. For example, if the number of harvesters in an area is restricted to give a sustainable offtake when the calculation is first made, this same number of harvesters could be taking substantially more a few years later, when they have all upgraded their hunting equipment.

- **Seasonal closures** are useful components of management if there are particular times when the hunted population is vulnerable (for example, around mating or births). They are not adequate in themselves though, as they don't directly reduce overall offtake levels. For example, even if the season is very short, people might respond with huge harvesting effort.
- **Spatial restrictions** are an ancient management method, dating back to the first royal hunting reserves and beyond. It is a fundamental component of our conservation heritage, as expressed in nature reserves and other types of protected area (Adams 2004). In recent years it has become widely recognised that spatial restrictions on hunting are potentially useful not just for conservation, but also in order to enhance harvest yields, through providing a reservoir of individuals to restock depleted areas. The idea is particularly current in discussions of marine reserves (Gell and Roberts 2003).

One of the main advantages of closing areas to harvesting is its **robustness to uncertainty**. If we are protecting a proportion of our stock, we can be more confident about instituting sustainable use schemes in other areas, knowing that we have a buffer against failure if our estimates of sustainable hunting levels turn out to be wrong. Instituting permanent or rotating closures of parts of a hunted area (also called **no-take zones**) is simple, transparent, relatively easy to enforce and has knock-on benefits for the rest of the ecosystem. However, there are still ongoing arguments about the degree to which spatial restrictions actually increase yields for the harvested area as a whole (Hilborn *et al.* 2004). This depends principally on how depleted the stocks are and the rate of dispersal from the hunted to the unhunted area. There are also questions about what proportion of the hunted area should be closed to optimise its effectiveness. The evidence from real systems about the successes of no-take zones as a harvest management measure is ambiguous (Willis *et al.* 2003), while from the conservation angle, protected areas have been criticised for their effects on the socio-economic status of local people (Borrini-Feyerabend *et al.* 2004). Thus, although spatial restrictions have a lot of advantages, there are also issues that need careful consideration. In particular, when working with local communities, a **participatory approach** is needed to ensure that zonation is done with the consent of all stakeholders (Beger *et al.* 2004; Box 7.4).

6.4.2 Promoting goodwill and cultural value

There is a huge range of interventions that fall under this heading, some of which also contribute to improving opportunities for alternative livelihoods. At the most basic level, just **talking to people** about your work, why you're in their area and what implications it may have for them is a basic courtesy that can lead on to fuller participation by the community in conservation and by conservationists in the community. Helping people in the communities where you work, being a good neighbour, is also an important component of conservation success (Section 3.2.3.4).

It's not possible to live and work in an area and interact with local people without becoming aware of their needs and priorities and wanting to help them to improve their lives. This can then lead to conservationists getting into more traditional development interventions such as provision of water, health care, school buildings, or taking an advocacy role on behalf of the people who they interact with. There are two major issues to consider when doing this:

- It is very important **not to raise people's expectations**. When outsiders come to an area, people wonder what impact this will have on their lives. A lone researcher into the sustainability of resource use cannot necessarily promise any immediate improvement to people's lives based on their work. The key is to be open and honest about the implications and the limits of the study or the conservation intervention for local people.
- It is important also to **remain objective** about what the intervention's aims are, and how best to achieve these. Both conservation and improving people's lives are important aims, and they can act synergistically, but it is easy to get distracted on the ground. Difficult issues about moral imperatives are common, and these need to be thought through, rather than ignored.

One of the commonest types of engagement with local people is through educational and **awareness-raising activities**. These may include leaflets, posters, calendars or t-shirts with a logo and conservation message on them. The medium depends on the locality—in Russia, for example, credit-card sized pocket calendars are very popular, whereas a t-shirt would be gratefully received (and worn) in Indonesia. These kinds of dissemination material are *de rigeur* in conservation projects nowadays, and they do clearly raise the profile of the project. But they probably have a negligible effect on people's actual behaviour (although the degree and direction of their influence remains to be quantified). Talks at schools, providing educational materials to children, holding video shows and if possible hosting trips into the field to see the conservation target are more fully engaged ways to raise the project's profile. Because they also involve interaction with people, and show concern for the area's children, they will give the project much more credibility than just distributing information passively.

Giving feedback to communities who have participated in research is really important to ensure that they feel involved and can see what the research has achieved and what the next steps should be. For example, a simple leaflet showing the overall results of an attitudinal survey in pictorial form can be distributed to the communities that took part in the survey. Alternatively a village meeting can be held to discuss the results, particularly if the community is predominately illiterate.

Ideas for awareness-raising and community participation in conservation are limited only by our imaginations. For example, a drama group made up of local unemployed youths was formed to deliver conservation education around Hwange National Park in Zimbabwe (WildCRU n.d.). **Participatory video** is another exciting new tool that could be transferred from development to conservation. This was used in Turkmenistan, where local herders themselves made films about their experiences since the break-up of the Soviet Union, with facilitation and training from outside (Insight n.d.). These videos were very influential with decision-makers at the national and international levels, and gave otherwise unheard people a voice in a popular and easily transferable medium. Invoking people's competitive spirit is often a good way to get them involved. The Durrell Wildlife Conservation Trust uses **quizzes and competitions** with small prizes to raise interest in conservation (Andrianandrasana *et al.* 2005; see also Box 7.3). Similarly Sirén *et al.* (2006) used a lottery as a vehicle for discussing the reasons why people hunt bushmeat.

6.4.3 Alternative livelihoods

A wide range of approaches to improving livelihood prospects from sources other than exploited species have been tried. What might work in a particular location needs to be assessed during the initial research phase, and should be based on the views of local people on feasibility and robustness (Chapter 3). Broadly, interventions fall into two types: obtaining better outputs from existing enterprises and starting up new enterprises. **Better outputs from existing enterprises** may include providing veterinary care to livestock; better breeds of livestock or crop; offering help in improving agricultural practices through, for example, water conservation and erosion mitigation; and capturing more of the value-added from a product via on-site processing (for example, cheese production, making handicrafts from wool, brewing). **New enterprises** might include natural resource based ones like honey production or weaving, or service provision such as bush taxis. Provision of micro-credit may open up a range of opportunities that were not previously available. The NGO Fauna and Flora International are pioneering a small grants programme as a method of kick-starting local businesses (Box 6.7).

Options for alternative livelihoods should be assessed on the basis of:

- Having been **proposed by local people** (otherwise they will not have ownership of the idea and it is likely not to be feasible or accepted). As facilitator you can help people come up with suggestions, but they must in the end feel that the idea is theirs.
- Having no negative **environmental impact**, particularly on the species or ecosystem of conservation concern. Ideally there should be positive linkages between the proposed activity and the environment, although this is usually indirect, through the enterprise contributing to the sustainability of development.
- If a primary aim is to improve the food security or wellbeing of the community, then **targeting women** is usually the best way to ensure the improved income is used for the benefit of the household as a whole.
- If the aim is to distract people from over-use of resources (see Section 6.3.3 for caveats about this approach) then the **resource users** must be targeted—and this may well be adult men. Hence there may be a lesser gain in terms of poverty reduction to trade off against the greater conservation gain. This is because additional resources gained by women tend to be invested into

improving the welfare of the household in general and children in particular, while additional resources gained by men are less reliably transferred (FAO 1996; Haddad *et al.* 1997).

- A clear understanding of why this alternative has not been adopted previously—what are the **barriers to adoption**, and are they surmountable? And once surmounted, is the activity likely to be self-sustaining? Surmountable barriers may include poor transport links to markets, lack of access to start-up capital or lack of training and expertise.

Box 6.7 Small grants for alternative livelihood activities.

Fauna and Flora International has pioneered a small grants programme (SGP) approach to promoting sustainable alternatives to natural resource use. The most established example of this is in Kyrgyzstan, where the programme has been running since 2000. The sequence of steps in implementing the programme is:

- Identify a local NGO to work with. This is an important part of local capacity-building, ensuring the project has a legacy in terms of an empowered NGO sector, and that there is local ownership and oversight. The NGO will probably need training in administering a small grants programme before the project starts.
- The local NGO and external partner hold a series of meetings in the target community, explaining the rationale behind the SGP, the criteria on which grant applications are judged, and discussing preliminary ideas about what kinds of enterprise might work. This ensures transparency about the aims of the SGP.
- Individuals within the community are asked to submit proposals to the local NGO for small grants (generally substantially less than $1000 each) for starting up or developing an enterprise. For example, the kinds of project proposals submitted might include setting up a mechanic's workshop, a honey-making enterprise (Figure 6.10), or a rug-making cooperative. This process ensures that the proposals are locally generated by individuals, rather than being externally imposed.
- The proposals are judged against transparent criteria by the NGO, an independent in-country steering committee and the external partner. These may include value for money, feasibility and contribution to environmental sustainability. Those proposals which look promising are taken forward, and the applicants work with the NGO to prepare proper budgets and a business plan. This gives applicants training in business management skills which help to empower them beyond the initial SGP.
- Proposals that are funded are then implemented, with the NGO supporting the recipients to make a success of their enterprise. The whole process is geared towards capacity-building rather than being a hand-out of money that is likely to be spent without taking the community any further forward. *Source*: http://www.fauna-flora.org/eurasia/kyrgyzstan.html.

Fig. 6.10 Bee-keeping enterprise, Kyrgyzstan. Photo © Juan Pablo Moreiras/Fauna & Flora International.

6.4.3.1 Tourism

Tourism is a form of direct but **non-consumptive use** of natural resources that can move people away from harvesting. It can be a useful method of bringing income into communities that is directly linked to the continued presence of charismatic species or ecosystems. Much has been written on the principles and practice of tourism as a conservation tool, mostly based around the concept of ecotourism. However, Kiss (2004) sounds a cautionary note about whether small-scale community-based ecotourism is a good solution to conservation problems, having many of the same issues as other methods of promoting alternative livelihoods. Another approach is to target the high-end luxury tourism market (Box 6.8).

6.4.4 Payment for conservation services

Latacz-Lohmann and Schilizzi (2005) have carried out a useful review of lessons from the implementation of one kind of direct payment for conservation, auctions for conservation contracts in agri-environment schemes. Many of their observations are more broadly applicable, however. One of the main issues to consider when implementing direct payments for conservation is whether to **pay by inputs or by results**. If you pay by results, then the link to conservation is obvious, but to people's efforts less so. In the face of external influences such as climate, pollution and disease, for example, the fact that people have reduced habitat degradation or hunting may not be reflected in changes in animal population sizes. This may be demotivating. Payment by inputs rewards effort, but may reduce people's creativity in trying to mitigate other causes of population decline (perhaps by influencing others to stop their destructive activities). If people have relatively full and predictable control over threatening factors then payment by results should work well, otherwise payment by inputs may be seen as fairer.

Another key issue is the **conservation index** that will determine the level of payment. This needs to be simple to monitor (see Chapter 7), relevant to the status of the conservation target, and objectively verifiable. Positive indices of conservation outcome might include some measure of animal population size or of change in status from the previous year (number of calling males recorded, perhaps, or number of occupied burrows). Or an index of harm averted through changes in activity might be the amount of habitat damaged or restored, number of trees cut in an area or number of poaching incidents recorded.

The **type of contract** involved is also an important consideration. In some places (e.g. Costa Rica; Zbinden and Lee 2005), a direct payments scheme is administered by the government. Landowners sign upfront legally binding contracts

---

Box 6.8 Maluane—high-end tourism with conservation at its heart.

Cabo Delgado in Mozambique is an area of high biodiversity and relatively low human impact, including healthy coral reefs and nesting turtles. The Cabo Delgado Biodiversity & Tourism project was started there in 1998, aiming to conserve the area through luxury tourism. The project has built up slowly and carefully, starting with research and monitoring, and opened its first lodges on the island of Vamizi in 2005, with substantial press interest (e.g. Ecott 2006; Figure 6.11). People on the island live a subsistence lifestyle based around fishing. The CDBTP is the only employer on the island. Among the many benefits of the project to villagers are employment, sale of fish to the lodges and access to training and social services. However, the project is facing external pressures, particularly substantial immigration of fishers whose own fishing grounds are depleted. This is potentially a major threat both to villagers' livelihoods and to the health of the reefs upon which both they and the project depend. There are two immigrant fishing camps on the island which have only started to grow since 2002—after 3 years, both had around 100 houses, and one had become a permanent settlement (Hill 2005). The CDBTP (now called Maluane, http://www.maluane.com/home.htm) has a way to go, but will be an important case study of whether high end tourism, building on a foundation of thorough scientific research and planning, and explicitly aiming to work in partnership with local communities, can succeed in conservation in the face of growing pressure on the ecosystem.

Sources: http://static.zsl.org/files/cabodelgado-4.pdf; Hill (2005).

Fig. 6.11 Vamizi lodge. Photo © Nick Hill.

for management over a certain period of time, which entitles them to payments. The International Snow Leopard Trust (Box 6.6) uses a price premium on goods, paid after the fact to individuals based on the performance of their community in averting poaching, measured as there being no poaching incidents recorded in the vicinity of the community. This is a fairly extreme form of payment by results (because the individual can only guarantee not to poach themselves and to try to influence others not to poach), but is accepted because the base price and quantity of goods to be purchased is guaranteed. Generally NGOs can only work through voluntary sign-up to schemes that threaten withdrawal of benefits should criteria not be met, while governments can institute legally binding land management contracts.

All conservation interventions need **monitoring** of outcomes, but it is particularly explicit in this approach, because the payments are dependent on reliable monitoring. Hence there is a need for simple and robust measures that can easily be measured, ideally by community members themselves to ensure transparency. It's also important that the scheme doesn't just attract those who wouldn't have caused environmental damage anyhow—it needs to be sufficiently attractive to people whose behaviour it would actually change. For example, in Lac Alaotra, Madagascar, in a scheme designed to conserve marshes for fishery enhancement and because they are habitat for the Alaotran gentle lemur, communities are banded according to the current status of their environment, and judged only against others in the same category. This provides more incentive for communities in poor environmental condition to take part, as they are being judged against those in similar circumstances (Andrianandrasana *et al.* 2005).

6.5 An integrated approach

Putting the components of a management plan in place requires an understanding of the system from the biological, social, political and economic angles, and of the reasons why management is required (Figure 6.12). This understanding will differ between stakeholders—local people may have a very different perception of both system dynamics and what the issues are to that of external researchers. The next step, then, is to build consensus between stakeholders through consultation, such that a workable solution can be proposed and a concrete plan of action developed. This solution is likely to combine aspects of several of the management approaches discussed above. After this, the management plan can be implemented, which is then followed by a continuous process of learning, adaptation and consultation. The actual management solutions that fall out of this process will be site-specific. Box 6.9 gives a couple of hypothetical examples.

Both the examples in Box 6.9 are relatively straightforward, at least in theory. The species are sedentary and the threats are local and controllable, unlike those affecting many other species—for example the Siberian tiger, which is widely dispersed at low density in remote areas where hunting is hard to police; polar bears, threatened by global warming; or the diffuse and informal bushmeat trade. The system is a manageable size—a national park or a village hunting catchment, and there is a distinct community that can buy into the management plan. In the first case the harvesting is already organised with an established commercial market, such that there is the prospect that a properly managed harvest could be

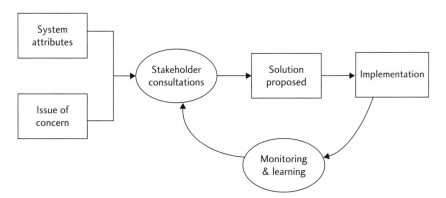

Fig. 6.12 Schematic representation of the development and implementation of a management plan. This is the basis for adaptive management, discussed further in chapter 7.

financially self-sustaining. In the second case, the species has no major livelihood value, and so the economic costs to hunters of ceasing to harvest it are minimal. But there are many more systems in which the approach to take is not nearly so obvious.

Box 6.9 Hypothetical examples of management plan development for conservation problems.

Case 1. A commercially harvested fruit

System: Commercially harvested high-value fruits of a plant with high resilience to exploitation but within a National Park. Major source of livelihoods to local people.

Issues: Park authority perceives a population decline, but the cause is unclear. Harvesters dispute that they are the cause. Authority also feels that presence of people in the Park is detrimental to the ecosystem. Users feel they are safeguarding the Park from outsiders.

Solution: Co-management of resource by community and Park Authority, requiring a change in legal (though not traditional) use rights. Community undertakes to monitor resource levels and control use. Stakeholders develop a set of access rules to safeguard the rest of the ecosystem from detrimental effects, and to monitor incursions by others.

Implementation: Collect data and develop an understanding of system dynamics, with harvester participation. Use this to develop an agreed quota-setting system. Cost-benefit analysis to ensure financial viability, particularly taking monitoring costs into account. Monitor progress and continue consultation to ensure adaptation.

Assumptions: Monitoring is practicable, and robust enough to detect trends and set a meaningful quota. Management and monitoring costs are low enough for new system to be profitable. Park authorities are prepared to devolve power and responsibility to local harvesters. Institutional structure is robust to changes in circumstances. External threats are containable. Cause of decline is correctly identified and addressed in the plan. Genuine adaptive management takes place. All stakeholders are committed to success.

Case 2. A recreationally hunted bird

System: Endangered species of little-known ground-dwelling bird, narrow range endemic. Opportunistic harvest for meat and recreation in the course of hunting other species, no commercial market.

Issues: Concern of conservationists that this offtake is driving species to extinction. Little local awareness of the issue. Popularity of recreational hunting.

Solution: An enforced harvest ban on this species, with other activities unaffected. Increased public awareness of the issue and cultural pride in their local species. Engaging local and international people in monitoring and research.

Implementation: Implement public awareness campaign to garner public support for conservation actions. Tie hunting ban to livelihood assistance for key hunters, for example, involvement in monitoring and research programmes. Develop a participatory monitoring programme to track changes in status.

Assumptions: Public awareness does lead to willingness to conserve the species. It is possible to discriminate between species when hunting without increasing hunting cost, and hence avoid killing this species. Opportunistic killing is the key threat to the species, now and in the future. Funds for local monitoring and research continue, rather than being short-term start-up only.

Note: Thanks to Lucy Rist and Kerry Waylen for the systems on which these examples are loosely based.

In the end, successful conservation comes down to getting the **incentives** right for all concerned. The **context** within which the conservation intervention takes place is crucial—what alternatives to wildlife use do people have? What are their opportunity costs of refraining from overexploiting their natural resources, what are the direct costs that they are incurring? The context extends up the governance chain—to what extent do local people have control over their natural resources and their land? What political will exists, or can be generated, at the local, national and international levels? Are corruption or inertia likely to sap the vitality of the project?

As we have seen in this chapter, incentives come in many forms. They can be positive or negative, and they can be use-based (providing people with economic benefits from conservation) or rely on promoting people's spiritual or cultural connections with nature. Some interventions, particularly ICDPs and projects that aim to distract people from wildlife exploitation, can lead to complex incentive structures for both individuals and communities. This may then give perverse signals that are not easy to anticipate, such as enabling people to invest in equipment that increases their hunting efficiency. Other projects rely on goodwill gestures, in which conservationists are not aiming to cover people's opportunity costs from conservation, but instead simply to demonstrate that they do care about the local community (Box 6.4). Direct positive incentives to conserve come in many forms. These might include bonuses to rangers (Box 6.2), benefits obtained through sustainable use of the resource (Box 6.3), or direct payments for conservation services rendered (Box 6.6). Negative incentives can be tangible, such as the probability of being captured and punished (Box 6.2), or more subtle through social disapproval and ostracism. The majority of mature and successful conservation interventions apply all these different types of incentives in a coherent package (e.g. Box 6.10).

In this chapter we have discussed the main types of management approach, their strengths and weaknesses, and how an understanding of the system and of the range of tools available can be combined to produce a management plan. The involvement of stakeholders (all those with an interest in the resource, not just 'experts') in developing this plan is the first stage in management success. In Chapter 7 we consider implementation in depth, with a particular emphasis on how to guide decision-making when there is uncertainty, and how to monitor the system and feed back the findings from monitoring into ongoing adaptive management.

Box 6.10 Community Markets for Conservation (COMACO).

Early experiments in integrating conservation and development have been widely criticised. However, experience has much improved the approach and now community-based natural resource management, one manifestation of this integrated way of thinking, is widespread. CBNRM has been taken up by governments around the world, particularly in southern Africa (see Resources section for websites). One particularly successful project is COMACO, which is based in the Luangwa Valley in Zambia and was initiated by the Wildlife Conservation Society. Although it has only been in existence since 2003, it is built on several decades of conservation experience in the area. It's also unusual in the clarity with which it states its underlying principles in the form of testable hypotheses.

COMACO's heart is a cooperative trading association for agricultural produce that guarantees fair prices at regional trading depots. However, its activities include tourism, wildlife law enforcement, forest management, capacity-building, promotion of sustainable farming methods and help for health professionals (Figure 6.13a). They also promote alternative livelihoods such as bee-keeping, poultry and fish ponds, particularly targeting ex-poachers (Fig. 6.13b). They process produce themselves to increase value-added, and to give a brand identity. This programme combines many of the strategies discussed in this chapter, including alternative livelihoods, generation of community goodwill and direct

(a)   (b)

Fig. 6.13 a) A course in conservation farming for COMACO members; b) a bee-keeper. Photos © COMACO.

use (through tourism). The ethos of the programme is market-based, and particular incentives to change lifestyle are directed at known wildlife poachers. There is also a strong development component through the targeting of help to food insecure families, leading to market opportunities that encourage farmers to remain committed to better farming practices, both for income and for food.

Source: www.itswild.org.

6.6 Resources

6.6.1 Websites

Community-based natural resource management links:

http://www.cbnrm.net/about/us/organisations.html
Online resources for NRM: http://www.frameweb.org/
Case studies in CBNRM: http://srdis.ciesin.columbia.edu/
COMACO, Zambia: http://www.itswild.org/
Namibia: http://www.met.gov.na/programmes/cbnrm/cbnrmHome.htm

Participatory video:

Insight: http://www.insightshare.org/
Maneno Mengi: http://www.maneno.net/pages/mmpv.html

Ecotourism:

The main industry organisation for ecotourism:
http://www.ecotourism.org/
UNEP ecotourism website, with useful links: http://www.uneptie.org/pc/tourism/ecotourism/
  home.htm
Ecotourism practitioners' forum: http://www.planeta.com/
Pro-poor tourism: http://www.propoortourism.org.uk/
Overseas Development Institute: http://www.odi.org.uk/rpeg/tourismpubs.html

Other:

Cochrane, K., ed. (2002) A fishery manager's guidebook: Management measures and their application. Food and Agriculture Organisation Technical Paper number 424. FAO, Rome. ftp://ftp.fao.org/docrep/fao/004/y3427e/y3427e00.pdf
Overseas Development Institute, Forestry Poverty and Environment Group: http://www.odifpeg.org.uk/
Paul Ferraro's directory of conservation payments initiatives from around the world: http://epp.gsu.edu/pferraro/special/special.htm

6.6.2 Textbooks

Caughley, G. and Gunn, A. (1996). *Conservation Biology in Theory and Practice*. Blackwell Science, Oxford. Full of examples of diagnosis and action for conservation problems, taking a biological and species-based approach.

Milner-Gulland, E.J., and Mace, R. (1998). *Conservation of Biological Resources*. Blackwell Science, Oxford. Covers the theory behind basic management approaches (such as harvest quotas) in more detail.

Sinclair, A.R.E., Fryxell, J., and Caughley, G. (2005) *Wildlife Ecology, Conservation and Management*. Blackwell Science, Oxford. On how to manage wildlife, from an ecological perspective.

Primack, R. (2006). *Essentials of Conservation Biology*, Fourth Edition. Sinauer, Sunderland, MA. A classic conservation biology textbook.

Implementing management for long-term sustainability

7.1 Scope of the chapter

In Chapter 6, we discussed a range of approaches to conservation management. Here we talk about how to implement conservation interventions in practice. We show how to draw together all the elements that we have discussed in previous chapters to build a robust framework for management. Research and management are mutually reinforcing, such that research results feed into management planning, but monitoring of the sustainability of our interventions deepens our scientific understanding of the system, which in its turn influences our management strategy.

We start the chapter by laying out some of the realities of conservation that people learn through experience but are rarely explicitly stated. We go on to consider setting clear objectives for your intervention, how to ensure that you can demonstrate whether you have succeeded or failed, and whether the output measures that are commonly used are really measuring the effectiveness of conservation interventions. This leads into a discussion of monitoring—the methods by which we can track progress towards sustainability. Monitoring requires the data collection techniques that we set out in Chapters 2 and 3, but we also need to consider pragmatically what is and is not feasible in the long term, given the financial and human resources available to us. We use the data we have collected to give us an understanding of the dynamics of the system, which enables us to make decisions. Here the analytical techniques we discuss in Chapters 4 and 5 come into play. We also need to consider that we as managers are part of the system, and there are new tools available to extend our models explicitly to include our management actions. Some of the management options that are available to choose from were discussed in Chapter 6. The final sections of the book remind us that conservation of exploited wildlife does not happen in a vacuum—there is a context of social, political and environmental change that impacts on our actions, and which we need to integrate into our planning.

7.2 Management in the real world

There are many texts available which give excellent advice on how to set up and manage conservation projects, emphasising inclusion of stakeholders, clear management goals and proper monitoring and evaluation of project success (see Section 7.8). It is easy to get a rosy picture of project implementation when reading these books in the comfort of your own office. But conservation is messy and difficult in the real world. It is also hard for conservation professionals to talk openly with examples from their own experience, because this can damage their ability to continue to work in an area. Some principles about how conservation works in the real world are given below:

- There is never a blank canvas—*pre-existing relationships* between individuals and institutions constrain possible actions.
- Everyone will want a *piece of the action*. Often the people who end up doing the work are not the ideal people for the job. They may have influence or just fit a particular profile (a particular nationality or employed by a particular organisation).
- Many projects are compromised by lack of *money*. However, if there is too much money this is also a danger, as it attracts corruption and mismanagement, particularly if there is pressure to spend to a timetable.
- People who understand and adapt to the *political, social and cultural* context of their work are far more likely to be successful than those who do not recognise the enormous socio-cultural divides that can exist between people. This particularly applies to highly educated people from developed countries who can be very rigid in their view of the world.
- For this reason, the most effective people are those who work *on the ground* in the long term, living with local people and speaking the language, often on small budgets, rather than flying in and out. They need to have input from outside though, to stop them losing perspective.
- Projects do not evolve at a consistent and measured pace. They happen in bursts of activity followed by stasis, and are often *implemented piece-meal* as funding and capacity are available, rather than in logical and tidy packages.
- Projects are implemented and assessed in a *short time frame* (1–5 years) and so short-term success is required to demonstrate value for money to donors. This mitigates against long-term sustainability, and pushes people towards easy wins.
- Funding goes to *fashionable* management strategies and conservation issues, that tick the right boxes with donors, not necessarily to the most effective management for the particular case. Similarly, investment levels and priority-setting are political and fashion-led.
- It may be partly true that individuals and organisations involved in conservation are idealistic and not primarily driven by the profit motive, but there are still strong incentives to attract funding, enhance individual careers and organisations' reputations that mean *people's motives* are not always as clearcut as might at first appear. Idealists can also be dogmatic and ruthless in pursuing their goals.
- Successful conservation involves making trade-offs, *compromises* and deals with influential people.

- Because all conservation involves many viewpoints, which rarely fully coincide, projects tend to need *fudged objectives* rather than clear ones, and to concentrate on low conflict outcomes rather than addressing the difficult issues.
- The truth about success and failure is rarely *openly reported*. There is no incentive to report failure.

So the question then is how to make the best of the circumstances in which a conservation project is embedded, to maximise the chance that you can make a useful contribution. Some suggestions are:

- There is no point in being rigid about your project implementation plans. You need to work within the system and be *flexible* and open-minded, without compromising your fundamental principles.
- *Take advice*, read the literature, and learn from others' mistakes.
- Be *reflective* about your own assumptions and behaviour, rather than passing judgement on others.
- Tread the *fine line* between speaking out strongly about what you believe in order to engender action (which can alienate people and potentially put the cause of conservation back several years), and compromising too readily (which can allow people to ignore the issues and not take the urgent action which is required).
- It is vital to identify and *fully involve* all stakeholders from the start. It is very dangerous to have interest groups or individuals around who feel disenfranchised, because they can work against the project.
- *Support* to the fullest people who you find to be honest and dedicated. Go out of your way to help them.
- Be aware that everyone has *hidden motives*.
- Never get involved in *corruption* or any other activities you feel uncomfortable with ethically.
- Insist on the importance of spending time reflecting and *analysing the problem* before management starts, rather than jumping in with management that people are pushing for. If a project starts off in the wrong direction it can be impossible to turn it around.
- Check that all activities are actually *cost-effective* and contribute to outcomes, and that all risks are properly considered in advance.
- Ensure that *long-term targets* are included from the start, for example, by setting up a simple, practicable monitoring system and by working to attract continuation funding from the start, to ensure continuing project viability at the end of the initial intervention.
- Build *local capacity* and be ready to take a back seat as soon as possible during the project period, so that when you withdraw, there is already confidence and experience available for project continuation.
- Bear in mind, however, that *long-term commitment* of funding and capacity is still likely to be required (i.e. 10 years or more), to ensure that the project does not disintegrate as soon as outside support is removed.

- *Stick to your strengths.* You don't have to do everything all at once. If your strength is in scientific research then that's fine, so long as you make it policy-relevant and engage with other people who will be using your work.
- Contribute to *reporting* and meta-analysis of conservation experience (through www.conservationevidence.com), however painful it may be if your experience is not of success.

7.3 Designing for success

The first step in designing a conservation intervention is to have some goals. The conservationist's overarching goal may be deceptively simple—to save species X from extinction. But this needs to be quantified. What do we mean by 'saved from extinction'? A population in a gene bank or zoo is probably not what we had in mind, but what is a viable wild population? How many individuals do we need, and how much habitat?

In a small-scale, short-term intervention, we may have a less ambitious goal; to stabilise the population of species X in area Y within timescale Z. We then need to quantify what we are going to do to achieve this goal, and how we are going to measure whether we have achieved it or not (Figure 7.1).

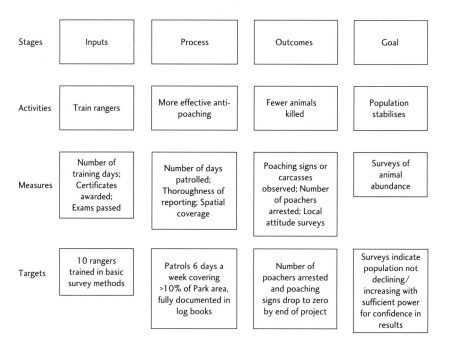

Fig. 7.1 The steps involved in conservation interventions, together with measurable targets illustrated for a project to improve law enforcement in a Protected Area.

In Figure 7.1 we see that each stage of the conservation intervention involves a set of activities, which can then be measured and compared against targets. Conservation funding bodies are looking more and more at ways to ensure value for money by measuring success against targets. One of the pioneers of this approach is the UK Government's Darwin Initiative, which publishes a list of output measures which all projects have to report against (Box 7.1). However, it is noticeable how concentrated these measures are at the left-hand side of Figure 7.1.

Box 7.1 The standard output measures for Darwin Initiative projects.

The Darwin Initiative is a small grants programme aimed at using the UK's biodiversity conservation expertise to help countries rich in biodiversity but poor in resources to achieve their commitments under the Convention on Biological Diversity. The outputs are used to measure the project's progress and final achievements.

Training outputs:

Number of PhDs, MScs or other qualifications obtained
Number of person-weeks of training given to students and others
Number of types of training material produced

Research outputs:

Number of weeks spent in host country by UK partners
Number of management plans produced
Number of field guides produced
Number of papers published
Number of databases and species reference collections produced

Dissemination outputs:

Number of conferences or workshops organised and attended
Number of press releases in UK and host country
Number of newsletters/dissemination networks
Number of appearances on TV, radio, newspapers in host country and elsewhere

Physical outputs:

Value of physical outputs handed over to host country
Number of permanent field sites/field plots/research facilities established

Financial outputs:

Amount of matching funding raised

Source: http://www.darwin.gov.uk/reporting/.

So why do conservation organisations typically measure achievements at the input or process stage, when we are actually interested in whether the project has been able to achieve its goal of conserving species? A key reason is that those carrying out projects know that they can reliably achieve their inputs within a short and specified timescale, and can show concrete proof of those inputs having occurred (Table 7.1). Inputs can't be so easily derailed by external events beyond the project's control, and so the project manager is within their comfort zone in promising to achieve them. True conservation achievement may well come from a planned set of inputs, but in reality it may also come from lucky breaks, changes of government, personal connections, favourable weather. Similarly a careful set of inputs may not lead to the required outputs due to economic or social collapse, political change, disease epidemics, etc. It is also likely to take longer than the typical lifespan of a project until we can be sure that the goal has been achieved. Narrower project outcomes may be the most we can hope to measure in time for the final report to sponsors.

This is not to say that inputs should not be measured, and targets should not be set for each stage in the process. Rather that it is important not to get hung up on inputs, without a clear idea of how these feed through to the conservation goal, and without thinking about how inputs may need to be altered to meet changing circumstances. The achievement of the conservation goal is the ultimate mark of conservation success. All other measures along the way are proxies for conservation success that primarily show the amount of effort put in.

Ensuring that projects are well thought through at the beginning involves making the linkage between goals and inputs explicit, as well as ensuring that the underlying assumptions are brought out into the open. 'Logframes' (logical frameworks) are one popular tool for doing this (Sartorius 1996). First used by USAID in the 1970s, they are now widely used in the development field, and are required by some conservation funders. They are useful tools for planning, but tend to be used as a one-off exercise in order to get money at the beginning of a project and an *aide memoire* when writing the final report, rather than as a tool for guiding project implementation. They also tend to impose quite a linear approach to what should be a dynamic and adaptive process of responding to threats and opportunities, and improving management based on experience. A good guide to how to produce a log frame is in Annex B of IFAD's online guide to monitoring and evaluation (IFAD n.d.; see Section 7.8).

Table 7.1 The measurability of different stages in a conservation intervention.

Stage	Timescale	Observability	Predictability	Impact	Comfort zone
Inputs	Short	High	High	Low	High
Process	Short-Med	High	High	Low	High
Outcomes	Medium	Medium	Medium	Medium	Medium
Goal	Long	Med-Low	Low	High	Low

7.4 Monitoring

It soon becomes clear that project design, data collection as part of the project, and monitoring and evaluation of project success are intertwined. At the project design stage, there needs to be some method included in the project plan by which data can be obtained to measure whether project outputs have been fulfilled. Project progress needs to be monitored against milestones, and evaluated to see whether it is adequate or not. Much of the recent interest in monitoring and evaluating in conservation has been concerned with these kinds of issues, ensuring that proper project management practices are adhered to and outcomes can be assessed (e.g. Margoluis and Salafsky 1998). This is vital, but it is not the only reason why we monitor in conservation. Monitoring is done:

- *To evaluate project success.* This is done at the end of the project, and asks how the project fulfilled its objectives, and what factors influenced success. If several projects are assessed, a comparative study can draw out general lessons (Salafsky *et al.* 2001).
- *To assess trends of conservation interest.* For example, the population size of the exploited species, the number of signs of human presence in a protected area, the composition of the offtake, changes in attitudes or incomes among users. This assessment of trends usually feeds into the evaluation of project success because it is measuring the results of the intervention.
- *To uncover and deter rule-breaking.* This is typically anti-poaching patrols.
- *To involve stakeholders in conservation.* If monitoring is done by or in conjunction with local people, either voluntarily or as project employees, this helps to ensure local communities feel ownership of the conservation intervention.

Monitoring for evaluation of conservation success should be part of project design, and is covered in Section 7.3. Monitoring to assess trends of conservation interest partly includes collection of the kinds of data covered in Chapters 2 and 3; for example, obtaining a series of estimates of population abundance or of local people's attitudes to nature, and relating these trends to conservation interventions. These data feed into evaluation of project success, and can be used to guide adaptive management (Section 7.5.2). However, both of these types of monitoring are prone to the 'project-based' mindset—the view that we have an intervention, which then ends and can be evaluated for success. Conservation cannot work in this way—it demands long-term involvement. Hence, important issues in monitoring for conservation are:

- How can we put in place a monitoring system that deters people from rule-breaking, and uncovers any violations that do occur?
- How can we design methods for monitoring trends of conservation interest that are simple, cheap and robust enough to last once the initial investment has gone, but that are still scientifically valid and have the power to detect trends before it is too late?

- How can we involve local people in monitoring, as a step towards locally based participatory management of natural resources, that in the long term is sustainable without external intervention?
- And the holy grail—can we combine the three points above, to get simple, cheap, scientifically robust, locally sustainable participatory monitoring that detects trends and at the same time promotes compliance?

7.4.1 Monitoring for compliance

In Chapter 6, we discussed the philosophical underpinnings and keys to success when setting and enforcing rules. Here we look at successful implementation of monitoring in order to ensure compliance with these rules. Enforcement and compliance with rules are emotive issues. The term 'poacher' is pejorative, and is not always appropriate, particularly when discussing local people using resources to which they feel they have rights. At the other extreme, poaching of high-value products is likely to be difficult to discuss openly with anyone involved, and offers the temptation for corruption among officials at all levels. Hence, this is a difficult subject about which to obtain validated and transparent information. It is also important to bear in mind that researchers, managers and local people can be placed in personal danger through persevering with enforcing rules.

In Protected Areas, monitoring is likely to be done by park rangers employed by the government, with sanctions applied based on legislation and through the courts. In community-based conservation initiatives, monitoring is usually done by the local people, and traditional systems of sanctions may be used. Alternatively, violators may face the withdrawal of privileges granted by the community or by an external NGO. In all cases there are some key principles for success:

- People respond to **perceived** risks of detection, capture and punishment (Box 3.2). This means that publicity is crucial whenever someone is caught, and that monitoring should be high profile. It is also no good detecting people if they subsequently go unpunished; this can be difficult to do anything about, because the authorities which administer punishment are often not the wildlife authorities, and may feel differently about the importance of the offence.
- As Jachmann and Billiouw (1997) demonstrated in the Luangwa Valley (Box 6.2), **following up leads** and working with local people to find out about rule-breaking can be more cost-effective than simply patrolling the protected area. The Wildlife Preservation Society of India has a similar approach to monitoring and publicising wildlife crime (WPSI n.d.).
- Monitoring need not necessarily be focused only on protecting species *in situ*. **Targeting bottlenecks** in the commodity chain, such as key traders, can be much more effective.
- **Monitoring must continue** even when poaching appears not to be occurring. If monitoring stops, then there is no incentive for people to continue abiding by the rules.

- Monitoring needs to be **dynamic and responsive** to poacher behaviour. If poachers go out in the winter, then rangers need to patrol then as well, however difficult the conditions. If rangers are following set routes around their bases, then poachers will soon learn to avoid these, and a false sense of security will ensue.

- Monitoring that leads to **informal sanctions** or withdrawal of privileges needs to be fairly and professionally carried out to ensure that resentment is not sparked by perceived arbitrariness.

- It is also important to **tailor monitoring** for compliance to the poaching threat involved. Locally based community monitors cannot be expected to tackle fully-armed elephant poachers.

- Monitors need to be well paid, trained, properly equipped and well supervised by a manager who acts upon their concerns and suggestions and is aware of the difficult social circumstances that they may find themselves in (for example, if friends or relations are implicated in poaching). This will enable them to do their job properly and take pride in their work. Monitors who are not properly supported are at best unlikely to be **motivated** to do their difficult job well, at worst will be tempted to get involved in poaching themselves, which they will be well placed to do given their specialist knowledge, equipment and access to the protected area.

7.4.1.1 Using law enforcement data to assess conservation effectiveness

One way to improve job satisfaction can be to include enforcement monitors in studies **monitoring trends** in biodiversity (Box 7.2). This gives a clear additional benefit to their patrols by collecting useful data for management, ensures that they see a point in continuing to monitor even when poaching is not occurring, and can enhance their professional satisfaction, particularly if they are involved in the analysis and interpretation of the data.

However, there are some caveats to this seemingly win–win situation. The routes that monitors need to follow in order to carry out law enforcement are non-random and should vary depending on poacher activity. By contrast, in order to detect trends in wildlife numbers, monitoring needs to be statistically rigorous and repeatable, such that observed trends from year to year are a true reflection of wildlife trends and not of changes in observer behaviour (Chapter 2). This is a major drawback which suggests that using data from law enforcement patrols to monitor wildlife trends needs to be very carefully thought-through to ensure biases are not introduced.

The interpretation of data on poaching incidents as a measure of poaching effort is also not as straightforward as it might appear. The underlying interaction between poacher behaviour and monitoring effort needs to be considered. The number of poaching signs observed by monitors is a non-linear function of monitoring effort, because there is an interaction between effort and compliance. If either effort increased but poaching activity stayed constant, or poaching activity increased and effort stayed constant, you would expect to see an increase in poaching signs (such as arrests or discovery of snares or damaged trees). However, because increased enforcement effort is not just passive monitoring but also leads to deterrence of poaching, you would expect to see low evidence of poaching activity at both high and low monitoring efforts, and high evidence of poaching activity at medium effort. Therefore, if there is no sign of poaching activity this can be because there is no effective monitoring, or because there is no poaching. It's not always obvious which of these pertains without independent information, because if monitors are not going out at the right times to the right places, it may appear that poaching has ceased.

In order to analyse monitoring data, it needs to be corrected for **observer effort**—for example, by creating an index of the number of poaching signs encountered per observer day. Over time, with consistent law enforcement effort which has the same probability of detecting any single poaching incident in each time period (i.e. there is no change in observer effort or in poacher avoidance behaviour), then a decrease in the number of poaching signs observed is good evidence that poaching has indeed declined. By dividing the number of signs by observer effort, we can look at changes in poaching activity over time even when the amount of effort put in is variable. But it is harder to correct for changes in the type of monitoring carried out and in poacher avoidance behaviour.

Box 7.2 Ranger-based monitoring for mountain gorilla conservation.

In the Virunga and Bwindi forest areas, the International Gorilla Conservation Programme has been involved in a ranger-based monitoring programme since 1997. Rangers collect data on locations of gorilla groups and poaching signs, and these data are visualised in maps and graphs of trends over time. The data collection has been useful in highlighting periods when incursions into the park to snare bushmeat or collect bamboo or water are particularly common; this tends to be in the dry season, June to August. Because the gorillas are habituated and tourist groups are taken in to see them, the data on gorilla locations has a practical use in guiding tourist visits and pinpointing areas of potential human-gorilla interaction, rather than being used to monitor trends in gorilla population sizes. Instead, periodic scientific censuses are used to give robust estimates of trends in population sizes. A comparison of information from household surveys (asking whether they collect water inside the park) and data on incursions from the patrols shows a strong spatial correlation, suggesting that the rangers are obtaining useful and verifiable information (Figure 7.2). A comparison of the location of human incursions and gorilla home ranges can be used to show the key areas of overlap and also areas where human presence may be excluding gorillas from otherwise suitable habitat.

Source: http://www.mountaingorillas.org/gallery/gallery_maps.htm.

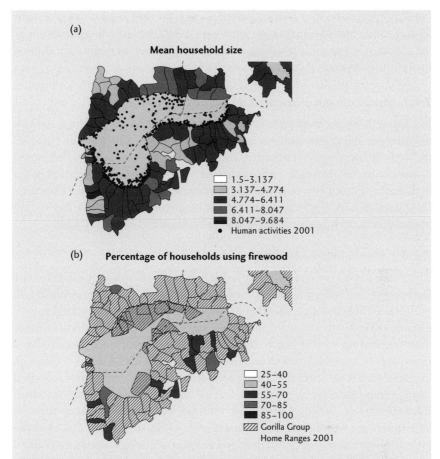

Fig. 7.2 Data collected by the Virunga National Park ranger monitoring programme. a) Dots show the location of human incursions into the Park, compared to mean household size in the vicinity. b) Location of gorilla group home ranges, compared to the percentage of households using firewood.

The next issue is whether a decline in poaching rate is due to effective deterrence through monitoring, other conservation activities (perhaps provision of alternative livelihoods or public education campaigns), or external forces (such as a reduction in demand for the poached good). This is not easy to ascertain, but is crucial for planning cost-effective conservation activities. Methods for disentangling these confounding effects include social research to discover individuals' views on the subject, or regression analyses attempting to correlate changes in poaching signs with changes in potentially causative factors (see Chapter 4). For example, a number of authors have used regression models to assess whether international legislation to limit the ivory trade under CITES has reduced elephant poaching (Dublin *et al.* 1994; Stiles 2004). CITES itself has set up an international programme, MIKE (Monitoring the Illegal Killing of Elephants; CITES, n.d.) to look into this.

Finally, there is a need to ascertain whether the reduction in poaching is actually leading to our desired conservation outcome—a stable or increasing population of the species of concern. If poaching is not the main cause of population decline, then effort invested in stamping it out will not be effective.

7.4.2 Monitoring ecological trends

In Chapter 2, we discussed methods for measuring population abundance, and how to determine the power monitoring programmes have to detect trends, based on the levels of sampling error obtained. In this section, we look at **cost-effectiveness** in monitoring, how to handle trade-offs between the cost of monitoring and its power to detect trends, and how to ensure that monitoring is sustainable in the long-term. Long-term sustainability in the context of small-scale community-based conservation implies that the techniques are simple, low tech, robust and easily analysable.

7.4.2.1 Participatory monitoring

The idea of participatory monitoring is that it should be bottom-up; carried out by local people, and collecting data of relevance to their management priorities (Halls *et al.* 2005). This should ensure that the monitoring is supported by local people, leads more readily to management action, and is substantially more cost-effective than monitoring carried out by or for outsiders. Participatory monitoring is immensely fashionable (see web resources in Section 7.8 for a few links), although in most cases it is used to monitor the success of interventions, rather than ecological trends. Some case studies do exist for its use to monitor ecological trends (Danielsen *et al.* 2005). The general message from these studies is that the approach is too new for a full assessment of its effectiveness, and that there are potential issues with the quality and bias of the data, meaning that inferences about ecological trends cannot always be made with confidence. However, there can be major benefits in terms of engaging and empowering local people. Innovative ideas have been generated to engage local people in monitoring (Box 7.3).

7.4.2.2 Using information from harvesting

Often we think of monitoring as a separate and independent activity from harvesting. However, harvesters traditionally use the information that they gain while harvesting as an indicator of population health. Catch per unit effort (CPUE) is a widely used indicator of population trends, the assumption being that as population sizes decline, individuals are harder to catch. There are a whole set of pitfalls which undermine the validity of this assumption (Chapters 2 and 4). Nonetheless, it is seductive to think that information collected while harvesting can help to inform management, and the potential benefits are large enough for the idea to be worth investigating. Using CPUE as an indicator of abundance forms a cornerstone of commercial fisheries management, because of the logistical difficulties of making an independent population estimate (Hilborn and Walters 1992). However, the information that harvesters use to assess the health of their resource

Box 7.3 Village competitions as a tool for participatory monitoring.

Durrell Wildlife Conservation Trust has been pioneering methods for engaging local people in conservation through participatory monitoring, making use of people's natural enjoyment of competitions. In the Alaotra wetlands in Madagascar, villages compete through monitoring key indicators of ecosystem health, such as observations of endemic species (particularly the critically endangered Alaotran gentle lemur, *Hapalemur alaotrensis*; Figure 7.3) and percentage of marsh burnt (a key component of habitat destruction). Villagers were particularly keen to include indicators of economic importance such as fish catches in their monitoring portfolios. The participatory monitoring has been going since 2001. Preliminary evidence suggests that trends in marsh burning are recorded well using this method, but that it is more difficult to get reliable trends in lemur population abundance. The method provides additional benefits in terms of community engagement with conservation. A similar scheme has been going since 2004 in the western dry forests of Madagascar, the only habitat of four threatened vertebrate species including the Malagasy Giant Jumping rat, the flat-tailed tortoise and the narrow-striped mongoose. Again, villages monitor key conservation attributes of their local area and compete for prizes based on indicators of conservation success such as numbers of occupied rat burrows.

Fig. 7.3 Alaotran gentle lemur at Lac Alaotra. Photo © Jonah Ratsimbazafy DWCT.

> The Durrell team highlight the initiatives' importance for ensuring transparency and good governance through open support of the scheme by village heads and politicians and improved acceptance of the presence of government officials in the area. Competitions promote a sense of pride in conservation and are ideal for public awareness and promoting media interest. It is too early to judge whether the monitoring is improving the conservation status of the target species.
>
> *Source*: Andrianandrasana *et al.* (2005).

is wider than just CPUE. Moller *et al.* (2004) give a table of types of information that harvesters use, which include animal condition, species composition and unusual events.

These types of information are part of **traditional ecological knowledge**, which is any information that is known informally about the environment where people live. Traditional ecological knowledge is a useful component of management, because by making use of it, managers can demonstrate that they respect users' expertise, and can enter into a dialogue about how best to manage the ecosystem (Aswani and Hamilton 2004; Drew 2005; Section 7.8). Combining local knowledge with scientific tools can be a very productive way to enter into co-management and build understanding (Box 7.4 Section 7.5.1).

Box 7.4 Including local knowledge and behaviour into a GIS for fisheries management.

The placement of Marine Protected Areas can be a contentious issue. Often the areas of highest fish species diversity and abundance are both key fishing grounds and prime targets for protection. GIS (Geographical Information System) technology is a particularly useful way of visualising local knowledge about fisheries, which is often spatial (where the best fishing grounds are at different times of year, where particular species aggregate or spawn). As well as being useful for combining datasets of various kinds, a GIS-based map is an interactive tool that can be used to help in negotiating decisions about MPA placement.

Aswani and Lauer (2006) worked with local resource users to map their knowledge of a reef system in the Solomon Islands, including where people fish, spawning grounds and substrates. The process of compiling this database was important in building local support for the MPA, and highlighting areas where the placement of an MPA would have caused major difficulties for local people. After the process, local support for the MPA was estimated at 70–90%.

GISs are difficult participatory tools. On the one hand, they are visually appealing ways to present and integrate spatial information. On the other hand, building a GIS is technically demanding, requires external facilitation and needs expensive equipment. So it is not an ideal tool for ongoing management at the local level.

There is a difficult balancing act to perform when using local knowledge in formulating and implementing conservation management. On the one hand, local people understand their environment better than outsiders, at least on their terms. Incorporating local knowledge can produce management strategies that are implementable, comprehensible and robust. On the other hand, and particularly when the species of conservation concern are cryptic, rare and not the main target of hunting, local knowledge may be lacking. Indices of abundance based on CPUE or body condition can be very misleading; one only has to look at the number of commercial fisheries collapses to appreciate this. It is tempting to suggest that these indices are adequate for trends in relative abundance, if not for absolute abundance, but this is not necessarily true either (Chapter 2). Local names for different kinds of plant and animal are another area of difficulty—they don't always overlap well with taxonomic species status, which is the important categorisation for conservation managers.

7.4.2.3 Cost-effectiveness

It is often assumed that monitoring ecological trends is an unalloyed good, and a fundamental component of conservation. While it is impossible to make good decisions from a position of ignorance, it is important to ask whether communities are likely to want to invest time and effort in monitoring trends in their resources (Box 7.5). It is also important to ask which type and level of monitoring will be most cost-effective (i.e. which will give the best trade-off between the power to

---

Box 7.5 Monitoring crayfish—is it worthwhile?

Hockley *et al.* (2005) developed a framework for evaluating whether it is in communities' interests to monitor their resources. They applied this to crayfish harvesting in Madagascar, using two types of monitoring—a mark-recapture study over 5 days, and an abundance index based on the number of crayfish caught in a single day. The abundance index produces population size estimates that are correlated to the mark-recapture estimates, but with far higher sampling error. Hockley *et al.* estimate that in order to have an 80% chance of detecting a 30% decline in the crayfish population, monitoring based on catch rates would have to involve 12 visits a year to each of 20 sites—a massive investment of time (720 person days per year). The mark-recapture method is much more sensitive—it is possible to achieve an 80% chance of detecting a 30% decline in the crayfish population by sampling once a year in 10 sites (50 person days). However, the mark-recapture estimate is substantially more complex to analyse, potentially beyond the capabilities of local monitors.

The study village contained 25 harvesting families earning on average US$0.76 per day from crayfish harvesting; a significant component of their livelihoods (Figure 7.4). However, given this level of income, it is highly unlikely

Fig. 7.4. Women selling crayfish. Photo © Julia Jones.

that they would be prepared to invest 720 person days per year in monitoring their resources that could otherwise be spent harvesting or in other economic activities. Loss of economically active time to monitoring could be avoided by using actual harvest rates instead of experimentally applied catches, but error in recording effort would make the results even less precise than the already rather weak 80% chance of detecting a substantial (30%) decline. These levels of power are fairly typical for wild populations, which may partly explain why local people do not indulge in detailed resource monitoring independent of harvesting. The issue for management in this case is to decide whether the low cost of harvest monitoring is sufficient reason to use it, despite its low power.

detect trends and costs of implementation). These questions have been asked in commercial fisheries science, because they involve investment in an economic resource (e.g. Clark and Kirkwood 1986), but are only just starting to be asked in conservation.

There are usually several methods available for monitoring ecological trends, and a number of studies are now appearing that compare them in terms of cost-effectiveness. Some studies discuss methods that are suitable for professional conservationists. For example, Joseph *et al.* (2006) look at when simple presence–absence surveys for birds outperform detailed counts, which give more information but are substantially more expensive. Gaidet-Drapier *et al.* (2006)

compare methods for counting wildlife in Zimbabwe, which range from the expensive and technically demanding (aerial surveys, vehicle surveys) to simple and potentially locally implementable methods (counting duikers on foot or from bicycles). The issues involved are that the faster you travel the more distance you cover, so the more individual observations you can make and so the more precise your estimate is likely to be (see Chapter 2). However, using a vehicle is costly and may create disturbance, and hence bias counts. Gaidet-Drapier et al. also counted animals at water points at the end of the dry season when water was very scarce—this is a cheap and simple method of getting an estimate of population size, but has major problems with bias (the animals coming to the water hole are an unknown proportion of the population, even if probably a high proportion at that time of the year) and is not amenable to statistical analysis. Weighing up the costs of sampling against the efficiency of detection, the authors showed that bicycle counts were the most cost-effective method for locally based monitoring.

To summarise the practical issues highlighted by these studies:

- It is crucial to think about **cost-effectiveness** when planning monitoring—if monitoring is not cost-effective it will not be sustainable.
- Cost-effectiveness includes an analysis of the **power to detect change**—cheap and simple monitoring may be a false economy because it produces data with high levels of error which can't inform management in a meaningful way (Katzner et al. 2007). It is important to do an analysis of the level and type of monitoring that is required to detect population trends, and then see whether this is realistic and affordable.
- If the level of monitoring that is required to detect trends is more time-consuming, expensive or technically difficult than the situation allows, then there is a real question as to whether it is worthwhile to invest time, money and community goodwill on it. This is where other solutions may be more appropriate. Rather than using monitoring to reduce uncertainty, it may be better to explicitly choose a management method that is **robust to uncertainty** such as setting aside no-take areas (Chapter 6). So an active decision is made to live with uncertainty about ecological trends, but to buffer against uncertainty by using a precautionary management method.

7.5 Making decisions

7.5.1 Who makes the decisions?

The structure of decision-making is fundamental to the outcome of exploitation (Table 7.2). At one extreme there are **open access** resources, with no overall control on how people use the resource. This lack of institutional control is the basis for Garrett Hardin's doom-ridden message that 'freedom in the commons brings ruin to all' (Hardin 1968). At the other extreme, the resource can be owned by an individual or by the state, who, in theory at least, can control completely the use that is made of it, and can manage for whatever objective they wish. The third type of

Table 7.2 Decision-making under different institutional structures.
Note: Those who make the decisions do not necessarily map either to those who own the resource or to those who use the resource. Common property, traditional use rights and community management overlap to some extent.

Institutional structure	Decision makers	Issues
Open access	Individual users	Prone to over-exploitation, users have no incentive to conserve.
Private property	Owner	Inequity of distribution, lack of social control over treatment of the resource. State may legislate to prevent particularly damaging uses.
State-controlled	State	Potentially a lack of accountability so local people have no say over use. Alternatively, lack of enforcement (hence state may not control use in practice, leading to de facto open access).
Common property	Collective, by a self-defined group	Not always robust to external influences, such as human population growth, market forces or imposition of other institutional structures on top.
Traditional/ customary use rights	Traditional users	Ownership can be unenforceable, prone to alienation of rights, often unacknowledged by outsiders.
Community management	Community, usually externally defined	Definition of community members and institutions often problematic, management may override rather than legalise customary rights.
Co-management	State + community	Degree of community involvement can be questionable, complexity of institutional arrangement, definition of community.

management is collective—when some group of people together makes decisions about management (Ostrom 1990). In a well-functioning democracy, there should not in theory be a line between state and collective decision-making—the state should be accountable to its people, and local concerns should feed up to national decision-makers. But in practice the difference is clear. There are a range of collective decision-making structures, which can crudely be categorised into **common property resources** (where the resource is held by a group of people), **traditional or customary use rights** (where use is regulated by local customs which may be unwritten and hard to codify), and **community management** (which is often relatively recent and has been set up with the involvement of the state). A new approach to natural resource management is **co-management**. This is where there is an explicit attempt to manage the resource as a joint venture

Box 7.6 Community forestry in Nepal.

Nepal has one of the longest-established natural resource co-management systems, based around community forestry. The system has been evolving since the 1970s, when in response to wide-spread degradation, the government introduced the Community Forestry Development Programme (CFDP). By 2002, about 25% of Nepal's forests had been handed over to 13,000 community forest groups, who represent about 35% of Nepal's population. This is no small-scale pilot project but a mainstream government policy which has had huge amounts of international donor support and interest. A community can apply to set up a forest user group (FUG) by submitting a plan for approval to the District Forestry Officer. The plan needs to include restrictions on access and spending of a proportion of the proceeds of timber sales on community development activities (about 36% in 2002). Although there are always issues associated with such a large-scale programme, particularly that the poorest are not getting the benefit that they should, the general view is that the programme has been successful, particularly in terms of improved forest extent and density. This qualifies as co-management because the FUGs manage the forest block on a day-to-day basis, but the Forestry Department provides the supportive legislative framework.

Source: Gilmour *et al*. (2004).

between the state and local people. It is particularly well advanced in fisheries (Arthur and Howard 2005; Pomeroy and Rivera-Guieb 2006), but has also been used for caribou management in Canada (Gunn 1998; Hurst 2004).

The whole ethos of participation in conservation and resource management suggests that **co-management** is a particularly good approach for implementing conservation in the modern world. In theory it should give the best of both worlds—the state involvement giving legitimacy, technical expertise, facilitation and access to funding, the community involvement giving local knowledge and the buy-in of the resource users. It is a natural follow-on from participatory research and monitoring—once communities are engaged, the logical next step is that they should be involved in ongoing management (e.g. Box 7.6).

All the issues that are raised about community-based conservation approaches and participatory monitoring in Chapter 6 and this chapter are also relevant to co-management. Some additional issues include:

- The concern that co-management covers a **range of community involvement**, from the communities being informed or consulted about management issues but having no formal power, right through to full community control but with the community being able to consult experts and obtain resources when they need them. There is the possibility that the term co-management can just dress up business as usual through the holding of a few consultations.

- If co-management is to work there is the inevitable need for another **layer of bureaucracy** with joint committees and structures for consultation. This is an additional expense and may also add inertia to management. The quality of the facilitation and leadership from both sides is vital for success.
- The community is working alongside a much more powerful institution. It is vital that the community is able to hold its own, which requires substantial **capacity building**, and a commitment to full participation from those representing the state. In particular the community may find it hard to get hold of information except via the state partner (for example, if there is technical expertise or infrastructure needed to get information about the status of the resource or patterns of use, see Box 7.4).
- The legislative position may be complex, and the **institutional structures** underlying co-management need to be well thought-through, so that each side has the intended level of control over management.
- As in all community-based conservation, there is the need to remember that a 'community' is made up of **individuals** and user groups with different, potentially conflicting, interests in the resource, all of whom need to be represented in the management structure.

7.5.2 How to make decisions

When we make decisions, there are a number of steps that we go through, whether consciously or not.

1. *Frame the problem.* What is the issue we are addressing? What are our measures of the success with which we have addressed it? What are all the possible options for action?
2. *Gather evidence.* What information do we have about the system?
3. *Model system dynamics.* Can we predict how the system will react to different management interventions? What is the uncertainty surrounding this prediction?
4. *Weigh up the options.* For each of our possible actions, how well do our predicted outcomes perform against the measures of success? This includes considering both our best guess at what the outcome will be and a measure of the range of possible outcomes and their likelihood under each action.
5. *Make the decision.* Based on the likely outcomes from each option, which is the best option to choose?
6. *Monitor and review.* The performance of the system once the decision is made gives us new evidence about system dynamics, and may also make us consider whether we need to reframe the problem. So we continue to gather evidence once the decision is made, and periodically go through the decision-making process again, armed with our new information.

This process is used subconsciously for our most banal decisions, such as the contents of the weekly shop, but it can also be formalised into **decision analysis**. Decision analysis is the process of quantitatively evaluating options under uncertainty. When decisions are complex and surrounded by uncertainty, and particularly when the issue is contentious, decision analysis helps us to lay out the steps taken in reaching a particular decision in a transparent way. This allows underlying assumptions to be exposed and challenged, and should help to ensure buy-in by all those involved.

Decision analysis is highly developed in fisheries science, and is now widely used to assess management options (Peterman (2004) and McAllister *et al.* (1999) both give excellent reviews). When used in fisheries science, it tends to involve sophisticated modelling, increasingly using Bayesian statistics. Cutting-edge decision analysis uses models and data to their fullest to give a robust understanding of system dynamics and the uncertainties involved. However, it is technically difficult and time-consuming, and it can be extremely hard to have full stakeholder involvement in the process when the modelling is this complex. Decision-making can still be significantly improved if the basic steps are followed, however, even if complex modelling is not possible.

The key advantages of this approach are that uncertainty is **explicitly included** at each step and that it separates value judgements about the desirability of different outcomes from people's understanding about the state of nature. It is important to make this distinction when there is a range of viewpoints and potential conflict about the approach to management. If disagreements about what the ideal outcome is are confounded with disagreements about how actions lead to outcomes, it is very difficult to reach common ground.

7.5.2.1 Framing the problem

The first thing to do is to ensure that the whole management process involves the right people. Generally, being inclusive is a good strategy, although practicality dictates that there is a limit to the number of people who can be involved. The best way to approach this is to carry out a stakeholder analysis (Grimble 1998). Who are the resource users, who has an interest in managing the resource? Who should be involved in gathering and interpreting the data on which the decisions will be made? Who should be involved in using this information to weigh up the options and make the final decisions? How can these groups be represented in the management activities?

Once the people involved have been identified, they can develop concrete outcomes and measures of success, as discussed in Section 7.3. Next, they can list the possibilities for management action that they wish to explore, perhaps based on the options discussed in Chapter 6, but also including the two extreme options of full protection and no action (business as usual).

7.5.2.2 Gathering evidence

The first step is to develop a conceptual framework for understanding system dynamics, of which Figure 7.5 is a simplified example.

We need to obtain and analyse data about each of the components of the system, using the techniques outlined in Chapters 2 and 3 and in the monitoring section of this chapter. This information can then be used to develop a model of system dynamics.

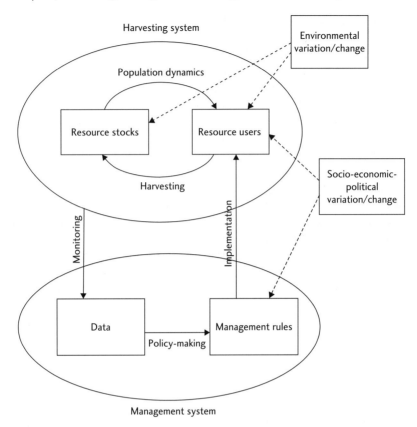

Fig. 7.5 A simplified representation of system dynamics. The crucial thing to note is that we cannot just consider the resource itself, nor even the resource and the harvesters—we need to understand the dynamics of management as well.

7.5.2.3 Modelling the system

Uncertainty

A fundamental component of modelling system dynamics is the way we treat **uncertainty.** Every component of Figure 7.5 is subject to uncertainty, and this needs to be quantified. People categorise uncertainty in many ways. However, the main types are:

- **Process uncertainty**. This is the most widely recognised type of uncertainty, and is generated by the fact that nature is inherently variable. There is variation in climate from year to year, chance events may raise or lower mortality or birth rates. If animals are not evenly distributed, people's harvest levels will vary depending on whether they come across a prey group or not. This type of uncertainty is reflected in the variability of the data collected in monitoring and can be addressed in data analysis using statistical techniques, and in modelling by giving parameter values (such as natural mortality rates) distributions rather than using point estimates.

- **Measurement uncertainty**. This is generated by the process of observation itself, and also includes the errors involved in generating parameter estimates from data. When we estimate animal population densities, for example, there are observation and sampling errors involved (Section 2.2.2). We may have crude instruments for measuring the number or weight of individuals harvested. When we interview harvesters about their offtakes, they may make mistakes in calculating numbers. This can be addressed by calibrating measurement instruments, modelling the error involved in estimation or by triangulating social surveys using a number of approaches (Section 3.2).

- **Structural uncertainty**. This is much more difficult to address, and more likely to be overlooked. When we build a conceptual model of how the world works, this model trammels the way in which we collect and interpret data. But it is not the only possible model of the world. When building models of population dynamics, for example, we need to assume a form of density dependence. But models using other types of density dependence may also be compatible with the data, and may give different predictions. When we consider how people might react to management rules, we are using a behavioural model. For example, we may assume that people act as rational profit-maximisers—which again may be misleading. The way to deal with structural uncertainty is to consider a range of plausible models of how the system might behave, and assign weights to them, based on the available data (Sections 4.4.1 and 7.5.2.5).

- **Implementation uncertainty**. Finally, there is the uncertainty about whether the decision made using this process will be based on the results of the analysis, whether it will be implemented effectively, and about the institutional context of the management itself. Some of this can be taken into account when considering structural uncertainty, but some is inevitably going to be outside the scope of the decision analysis.

A particularly good way of considering all these sources of uncertainty within a single framework is the **operating model** approach (Kirkwood and Smith 1996; Kell *et al.* 2005). This involves modelling both the management and harvesting systems, and explicitly including the observation process (Fig. 7.5). The complexity of the modelling necessitates a simulation approach, where the model mimics the main processes occurring in reality. A similar approach in ecology is the '**virtual ecologist**', a simulation of data collection that allows measurement error to be explicitly included in predictions (e.g. Tyre *et al.* 2001). One of the earliest and most influential applications of the operating model approach was by the Scientific Committee of the International Whaling Commission (Box 7.7).

Models for management

Although complex and sophisticated models are at the heart of decision analysis for fisheries, it is still worthwhile to use simpler models if complex models are not feasible. At the other end of the scale are simple diagrams of people's understanding of how the system works, which are informed by data in only a qualitative way, and can be generated at stakeholder meetings. As we outlined in Chapter 5, our view is that management is best served by employing a quantitative framework, including formal modelling based on data collected in a statistically robust and rigorous way. But the process of decision analysis is worth proceeding with regardless of whether formal modelling is possible.

7.5.2.4 Weighing up the options

Once the model of the system has been developed, the next step is to use it to evaluate the likely outcomes of different management actions. This evaluation needs to include some kind of robustness testing, in order to see how management strategies cope when things go wrong (Box 7.7 and Section 5.3.4). The first step is to choose a quantitative method for evaluation of the performance of management strategies. There are three components to performance, which should reflect the objectives decided upon in Section 7.5.2.1:

- A **conservation constraint**, representing ecological sustainability. For example, the population size of the exploited stock must not drop below a certain size. This is usually taken as the over-riding objective, which must be fulfilled first.
- Some measure of the **overall profitability** of the exploitation, based on the requirements of the resource users, representing financial sustainability. This may be the catch per unit effort in a subsistence system, or the monetary profits in a commercial system.
- Some measure of the **stability of the system**, representing social sustainability. This might include the equitability of distribution of the profits, livelihood security of the users, stakeholder attitudes towards management or compliance with rules.

Next we need a set of scenarios for evaluation of strategy performance. For example, how would each of our proposed management actions perform under the following circumstances?

- Our best guess at the system dynamics, using the data that we currently have.
- A crisis—either biological (a crash in prey numbers due to a bad winter or disease) or social (a major immigration, a change in hunting efficiency).
- A deteriorating situation (climate change-induced drought, or gradually increasing hunter numbers).
- A major failure in our understanding of the system (the population dynamics are completely different, the size of the population is overestimated, the population is aggregated in space, rather than being randomly distributed, weapons are much more efficient than realised, people are selecting their prey differently).
- Implementation is not as we might hope (low compliance with management rules, quotas actually set are higher than the decision-makers recommend).

Box 7.7 Innovative modelling for whales.

The Scientific Committee of the International Whaling Commission developed an operating model for use in determining precautionary harvest limits should commercial whaling recommence. They started by defining their objectives. There was an over-riding constraint that for harvesting to occur, there should be an acceptably low risk of whale numbers falling below 54% of carrying capacity. Once this constraint was fulfilled, managers could maximise yield, but with variation in yield remaining relatively low.

The operating model includes a number of submodels. There is the population model that generates the 'true' stock size, an observation model that transforms the true stock size into the observed stock size, and then a management strategy that determines the number of animals to be harvested based on the observed stock size.

Five candidate management strategies were developed and tested in simulation trials. These included base case trials, using best estimates of the state of the system, and robustness trials, which were sensitivity analyses showing how the strategies performed in a range of situations. The situations included biases in the catch data, incorrect assumptions about the biology of the stock, incorrect starting population

Fig. 7.6 Whale carcass, Hvalfjordur whaling station, Iceland. This picture was taken in 1983, when Iceland was still a member of the IWC. Photo © Michael Moore, Woods Hole Oceanographic Institution.

sizes, long periods of no data (due to the stock being protected and unhunted) before management began, catastrophes (e.g. epidemics) and deterioration in the environment. The most robust strategies should be able to respond appropriately to these situations by changing their catch limits so that true stock sizes do not fall below the threshold, but still be able to maximise yields (with low annual variability) if the stock is healthy. The simulation approach demonstrated that strategies that appeared to be precautionary and based on our best understanding of the situation did not always perform well when their underlying assumptions were not met—so it is always a good idea to test strategies thoroughly against a range of scenarios. One key feature of the winning strategy was that the harvest quota responded to the precision of the data available—if the data were more uncertain, the catch levels were lowered.

This strategy was adopted by the IWC. It is here that implementation uncertainty came into play—despite all the testing and the likelihood that this management strategy would enable whaling to be sustainable, political and welfare considerations meant that the strategy was not adopted. The IWC currently still has a whaling moratorium, and some countries are still whaling outside of the international regulatory framework (Figure 7.6).

Source: Kirkwood and Smith (1996).

Of course if a management action is tested in impossible circumstances it will fail. The idea of this exercise is not so much to demonstrate this as to give a feel for the range of possible outcomes. If this is done quantitatively, it will include a weighting of models by their fit to the data, which can then give a probability distribution of possible outcomes under each scenario (e.g. Hilborn and Mangel 1997; Box 4.5). This is likely to give a different result to just using best estimates, and one that is a better reflection of our true state of knowledge.

7.5.2.5 Making the decision

The result of the evaluation of the different actions is a performance index for each action against each criterion. It is relatively easy to rank actions according to their performance against each indicator. However, the criteria for success are not generally measured using the same units, and so there needs to be a way of deciding the weight to be given to each. For example, how do you determine the relative weight to be put upon the effectiveness of a management action in improving the livelihoods of poor families as against its ability to maintain a minimum population of the exploited species? If criteria are expressed as constraints (such as population size of prey at least 50% of carrying capacity, a basic level of offtake guaranteed for all) they are much easier to include in weighting exercises than those expressed as continuous variables (such as maximise yield, minimise variability). Either a management action fulfils the constraint or it doesn't, and if the basic criteria absolutely must be met, there is no requirement to weight or trade-off one outcome against another.

It is here that the views of stakeholders are paramount, and also that the time and effort spent in getting agreement on clear and realistic objectives and indicators at the beginning of the process will pay off. Although performance against constraints is easy to evaluate it is no use having constraints that are virtually incompatible with each other. If the aim of management is to make millionaires out of all members of the village while ensuring a viable elephant population, it may be easy to state but it will fail. Assigning weights to livelihood benefits against animal population sizes is much harder, but more realistic. Usually a combination of constraints and continuous variables is used, as is the case for the model in Box 7.7.

Conservation organisations have embraced the **precautionary principle** as a way to evaluate management actions (see Section 7.8). It is now enshrined in international law, including the Convention on Biological Diversity. The basic implications of the principle are that:

- Uncertainty should not be used as an excuse for inaction when there is a non-negligible risk of severe environmental damage from an activity.
- In the presence of uncertainty, the benefit of the doubt should be given to the conservation of the species or ecosystem under threat, rather than to the damaging activity.

Although it is a useful statement in international law, it is unclear how much further forward the principle takes us in making management decisions. It is open to much ambiguity and debate, allowing some to take radical stances that all potentially environmentally damaging activities should be prohibited, and others to discuss at length the semantics of what is non-negligible, what is severe damage, and what does giving the benefit of the doubt mean in practice.

As discussed in Section 7.5.2.3, there are now ways in which uncertainty can be explicitly included in models of system dynamics and in management planning, so the first element of the precautionary principle should be relatively uncontentious. The issue of which interest group should have the benefit of the doubt is more difficult, and is most case-specific. It may be easy to agree that, when a species is being harvested for commercial gain, even a relatively low probability of extinction from this activity might trigger a moratorium on this harvesting. We could quantify this by asking, as was done by the IWC (Box 7.7), what the probability of dropping below a predetermined threshold was in a given period, and setting an acceptable limit (for example, a 5% probability of the stock going below 50% of carrying capacity in 100 years). But if a species is relatively abundant elsewhere and local people depend on it for their livelihoods, as is the case for much bushmeat harvesting, why should the precautionary principle be used to prevent exploitation? Perhaps instead, the benefit of the doubt should be with the harvesters (so that, for example, any extinction risk of < 50% in 10 years is acceptable). Generally, the precautionary principle is best left in international conventions, and simply borne in mind when developing site-specific management objectives.

7.5.2.6 Monitoring and review

Although it is not realistic to carry out full decision analyses every year, it is important to remember that management is an ongoing and dynamic process which is put in place for the long term, and not just implemented in a 3-year project and then left to run itself. Management actions involve setting up the institutional structures that not only implement the chosen activity but also carry out monitoring and enforcement, and are able to use the results of monitoring to respond to changes.

Roughgarden and Smith (1996) proposed the 'rule of three-quarters'—that the best form of management is to determine the offtake that represents a stock size of three-quarters of carrying capacity, and then implement this in perpetuity. This is the antithesis of **adaptive management**, which has become a major theme in resource management (Walters 1986; Parma *et al.* 1998; Lee 1999). There are also shades of management approach between these extremes (Table 7.3).

There is a tempting simplicity to static management. If a no-take area is set aside, for example, and people can harvest in the rest of the area, it may appear that the problem is solved. However, no conservation solution is sustainable in the absence of monitoring for compliance, and there is also a need for ongoing ecological monitoring to ensure that the management outcomes are being fulfilled. Even this is not enough. In the real world, the context for conservation is always changing—the

Table 7.3 Approaches to management.

Name	Approach
Static management	Determine sustainable offtake, devise harvesting rule, monitor and enforce. No feedback between harvesting rule and outcome. Simple to set and enforce but not likely to be robust to changes.
Reactive management	Fire-fighting. Set management rule, monitor outcomes, change if things go wrong. Inefficient, may be missing better options. Response may be too late.
Passive adaptive management	Learn from outcomes of previous management actions. Feedback between outcomes and actions through review and revision. May not obtain maximum benefits through lack of exploration of options.
Active adaptive management	Management has learning as an explicit objective ('learning by doing'). Set up as an experiment to reduce uncertainty and explore system dynamics. Learning has short-term costs for potential long-term benefits. May be unsuitable for endangered species because varying management may be risky.

Source: Adapted from Shea *et al.* 2002.

community may grow or a new employment opportunity arise, the price for the harvested species may change, global climate change may erode the ecological value of the site, alien species may invade, etc.

If management is to be able to respond effectively to these inevitable dynamic processes, it has to be adaptive. That is, there needs to be **feedback** between monitoring and management action, so that as information about the system is collected it is used to improve management. This should be done at two scales. First, year on year, new data will be collected which can improve our understanding of the system and reduce uncertainty about how it works. This should be automatically included in the decision-making process for the next year, when quotas are set or licences granted (Box 7.8). Second, periodically there needs to be a full appraisal of the system, following the steps outlined in Section 7.5.2, so that the managers can re-evaluate their understanding of the system as a whole and possibly revamp the management plan.

Active adaptive management is much more radical. The idea is that managers should actively set out to **learn** about the system, rather than reviewing and revising based on what they observe. Learning about the system may involve having different harvest levels in different parts of the area, which might give information about the functional form of density dependence and about the linearity of catch

---

Box 7.8 Adaptive Harvesting Management for ducks in the United States.

Since 1995, the USFWS have been using adaptive management in their setting of harvest quotas for duck-hunting. They have a set of possible regulatory tools, such as changes in season length and numbers that can be shot on a particular occasion. They also have a set of models which describe duck population dynamics. Because they are not certain about key features of duck ecology, there are several alternative models, which are weighted according to the level of support that data from previous years give them. Each year, new data become available on duck population sizes and on environmental conditions from population monitoring and hunter surveys. These data are entered into the models, and a range of regulatory options for that year is evaluated. The best option is chosen based on their management objective (to maximise the duck harvest in the long term). Then monitoring of the harvest is carried out, these data are entered into the population models, and the cycle begins again.

This is adaptive management because the models are learning from year to year based on the new data that become available. But it is passive adaptive management, because there is no active attempt to experiment in order to learn more about the system.

Source: U.S. Fish and Wildlife Service, http://www.fws.gov/migratorybirds/mgmt/AHM/AHM-intro.htm.

per unit effort for the species. It might include changing the selectivity of gear types and seeing how that affects the size and composition of the catch. A control area might be harvested under business as usual to address the confounding effects of climate or ecosystem change on the outcome of harvesting. The major structural uncertainties in our models of how our system works can then be tested and reduced, allowing much better management.

However, active adaptive management is not easy to implement, particularly when conserving exploited species in poorer countries. Learning has a cost in terms of the yields foregone while testing out the system, rather than going straight for the hypothesised best management strategy. It may be risky, because the aim is to push the system into situations where it might not otherwise go, which could include very heavy harvesting risking population collapse or changes in harvesting strategy risking livelihood collapse. It may be hard to get stakeholders on board, given that the approach is relatively scientific and may seem rather divorced from the day-to-day concerns of resource users. There is always going to be a trade-off for managers between the scope of change that is politically and socially feasible and the potential benefits that could be gained from improved system performance. Finally, in rapidly evolving systems, an adaptive approach may fail because the rate of learning is slower than the rate of change; a simple precautionary approach (such as a no-take area) may be more robust in these situations.

Generally, active adaptive management has been much talked about but little implemented outside developed country fisheries (Lee 1999). One example in a developing country is the use of adaptive learning by stakeholders in deciding on stocking rates for aquaculture ponds in south-east Asia (Garaway and Arthur 2004), which is an exciting new development, but is in a context which is much easier to control than harvesting wild resources.

7.5.2.7 The scope of decision analysis

Decision analysis is a way to formalise management decision-making so that the process is transparent. However, while a logical and quantitative approach to decision-making is powerful, the approach will still not produce a magic answer to difficult problems. Weighting of outcomes will always be subjective and results are still limited by our understanding of the system. It is important that the quantitative nature of the approach doesn't give false confidence. In the end it is only by monitoring outcomes and adapting to circumstances, and by involving all stakeholders in the process, that management can be effective.

7.6 Contextualising management

It's easy to get caught up in the local and short-term detail of implementing conservation management. However, local issues are embedded within a wider context, which can derail the best conservation actions if it is not taken into account.

7.6.1 External factors affecting conservation success

7.6.1.1 Ecological issues

A key ecological issue is **climate change**. Is the conservation programme in an area which is particularly likely to be affected by climate change? Is it possible to plan to adapt to climate-related changes? If the answer to the first question is yes, and the second no, then perhaps the conservation intervention is not viable in the longer term, and needs to be reconsidered. Other external threats include **invasive species**. These can disrupt sustainable conservation planning. For example, in Biligiri Rangaswamy Temple Wildlife Sanctuary, India, an invasive mistletoe parasite is damaging wild fruit trees, threatening the livelihoods of local harvesters (Sinha and Bawa 2002). **Pollution** can have a significant impact on conservation success, and yet be diffuse and hard to deal with if the conservation intervention has a limited remit. Finally, **land use** outside the management area can have an impact on the success of conservation—for example, siltation and eutrophication from cultivation on land damaging coral reef ecosystems, or deforestation in the mountains affecting the dynamics of whole watersheds. This needs to be dealt with by expanding the focus of the intervention to include the area of the damaging activity—which may not easily be achieved.

7.6.1.2 Institutional issues

In order for conservation interventions to succeed, the **governance** regime needs to be robust. Governance is the act of governing, of running a country, project or organisation. It includes not only the government, but all who have power and influence, from individual people to organisations. The **institutional context** of an intervention (the organisations which have power to affect outcomes—local people, central government, international treaties, NGOs) determines the governance regime it is under.

A common pattern in conservation, particularly in developing countries, is that part of the reason for the conservation problem in the first place is poor governance and a weak institutional context. Perhaps a National Park is under heavy poaching pressure because park management is underfunded and has low morale, so cannot enforce the law. Or the lack of an ownership regime allows itinerant fishermen to come into a region and over-exploit fish that once sustained local villagers. Or corruption means that there is no attempt by the authorities to clamp down on illegal timber extraction from an area. Then an international NGO gets concerned and comes into the area, and this intervention improves the situation for a short time, because the NGO takes on the governance role and sets up institutions for resource management. It's as if the NGO is providing a '**governance bubble**' that protects the particular area from the prevailing conditions for the duration of their project. Then the NGO leaves, and fairly quickly the new local governance regime succumbs to the underlying institutional problems that had been there all along. Conservation-minded local officials are replaced, or management practices that are not immediately profitable or require technical skills gradually degenerate. The situation is soon back to where it was before the intervention started.

Tackling this problem is the reasoning behind many international donors' recent strategy to prioritise **capacity-building** in government administrations, run money through national institutional structures rather than directly to independent projects, and fund projects that national government sees as priorities. For example, much aid is now run through the 'Poverty Reduction Strategy Paper' approach (Piron and Evans 2004). These documents are prepared by the governments of countries receiving aid, and are meant to reflect their development needs and priorities. This matters for conservation and sustainable use, because rural development and poverty reduction are intertwined with sustainable resource use (Davies and Brown in press), and because this funding approach is likely to become more wide-spread. It can be frustrating for conservationists, though, because conservation usually comes low on government priority lists. Often small-scale independent interventions can be much more successful, in the short term at least, than bigger projects run through the government administration, especially if there is significant corruption.

The institutional context of a management intervention is a fundamental determinant of its long-term sustainability, and interacting effectively with the existing governance structure (be it local administrators, tribal chiefs or the responsible Ministry) is part of ensuring success.

7.6.1.3 External trends

Small-scale conservation interventions are often fighting to hold back the tide. As human population pressures increase, the sustainable output per individual from wildlife declines, and wildlife habitat is eaten up for agriculture. Trends for increased urbanisation, infrastructural development and international demand for goods affect communities and natural habitat. Projections for bushmeat consumption suggest that with current human population growth rates in the Congo basin, there will be a major protein deficit in the region and a collapse in populations of bushmeat species within the next 50 years (Fa *et al.* 2003). The Amazon basin is under pressure from the major infrastructural project of paving the Trans-Amazon Highway, which is likely to allow substantial increases in colonisation rate by subsistence farmers, and also from European demand for non-GM soya, which has allowed Brazil to take trade from the USA and become a major international soya producer (Laurance *et al.* 2004). There is little point trying to shield a small area from these powerful processes—instead conservationists must fight to be part of the wider planning process, so that environmental concerns are part of mainstream decision-making, rather than being seen as side-issues to development.

7.6.2 Cross-sectoral environmental planning

The only way to tackle the external threats to conservation action discussed in Section 7.6.1 is to internalise them. This requires conservation to be treated as a component of planning across all governance sectors. Rather than being considered as the job of the Ministry for the Environment, conservation needs to be thought about by government departments for trade, industry and finance, and by international conventions and donors of aid and development money. It needs to be a top priority when governments put together their Poverty Reduction Strategy Papers and when the World Bank decides which infrastructure development priorities are funded. Easily said, but less easily put into practice.

One useful step towards this is for governments to implement conservation at the **landscape scale**, allocating areas for different uses with conservation priorities in mind. Policy to address bushmeat over-exploitation, for example, can distinguish some areas where no hunting should be allowed (protected areas), some where hunting is for local use only and controlled by local people, and others where the government may have a say in maximising commercial yields, whether it be of timber or bushmeat (Bennett *et al.* 2007). One place where landscape-level conservation planning has been implemented is the Cape Floristic Region (Box 7.9).

Box 7.9 The Cape Floristic Region—a pioneer in conservation planning.

The Cape Floristic Region is a centre of astonishing plant endemicity, containing 3% of the world's plant species, of which 70% are found nowhere else (Figure 7.7). The fynbos ecosystem is home to many of these species. The Cape

Fig. 7.7 King Protea, *Protea cynaroides*, at Maclears Beacon, Cape Province. Photo © Nigel Forshaw.

Action Plan for the Environment was launched in 1998, and aimed to carry out systematic conservation planning for the region, identifying key areas that required protection, promote sustainable use of the wild plants and improve conservation implementation.

The CAPE process was academically driven, with much research based in universities aimed at correctly identifying key conservation areas. Less progress was made in engaging stakeholders at the right level, particularly local councils. However, the initiative has had a snowball effect, with big donors such as the Critical Ecosystems Partnership Fund (CEPF) coming in and focusing more on involving civil society in the process. The whole process has been successful not least in publicising the opportunities that exist to take a large-scale approach to conservation planning that is both scientifically rigorous and possible to implement in the real world.

Sources: Cowling and Pressey (2003), CEPF (2005).

7.7 A last word

Now is an exciting time to work in conservation. The issue of biodiversity loss is high on the political agenda of most countries, albeit mostly in the context of climate change rather than urgent short-term threats like over-exploitation and habitat destruction. The conservation community has a more nuanced approach to conservation action nowadays, recognising that polarised debates concerning the relative merits of stereotyped fences-and-fines and integrated conservation and development projects are unproductive. Instead, we are moving into a period in which the emphasis is on acting at the appropriate temporal and spatial **scale** for the problem at hand, and on **inclusive** action. This means fully involving all stakeholders, local people, government, researchers, and all types of expertise, be it traditional knowledge, biology, anthropology or social science. Rather than relying on case-by-case analyses, a **body of knowledge** is building up, allowing generalisations to be made about what types of intervention work best where.

There are still some gaps in our understanding, and in our ability to put this understanding into practice. As conservation philosophies come in and out of fashion, we need to remind ourselves that a range of approaches to conservation is available. Focusing on one approach to the detriment of others is never a sensible strategy. For example, over-reliance on the 'use-it-or-lose-it' philosophy of assigning economic value to endangered species risks undervaluing the important role of **cultural and spiritual values** in motivating people to conserve their natural heritage. Focusing on the rights, needs and aspirations of people living with wildlife has to be the right way forward—but we also need to remember that as conservationists we are trying to preserve biodiversity for **wider society**, which may involve restricting individual freedom to exploit wildlife.

In terms of conservation science, there is still a huge gap in methodology and approach between those who use sophisticated science to manage commercially important resources (fisheries, timber and game) and those who conserve endangered species. This gap must be bridged, for the sake of both sides. In fisheries science, decision analysis based on Bayesian modelling, and leading into **adaptive resource management** with stakeholder participation, is becoming standard best practice. But this approach needs to be translated into appropriate tools for conservation management, where both technical and financial capacity are much lower. In some systems, such as waterfowl conservation in the USA (Box 7.8), this is beginning to happen. Natural resource managers, in their turn, could benefit from the expertise that many conservation scientists have in building socially sustainable systems from the bottom-up, with the support of resource users and the public at large.

Development approaches were at first applied to conservation problems in a rather unsubtle way, such that conservation aims were sometimes forgotten. However, a new generation of conservation and development projects is emerging, like COMACO (Box 6.10) or Maluane (Box 6.8), in which **economically successful enterprises** are being set up explicitly to support both conservation and development goals. It is notable that although both these examples are recently set up, they build on many years of research and consensus-building.

An emerging theme is the importance of ensuring that natural resource conservation is not seen in isolation from other sectors of the economy, but as a key part of developing a country's social and financial wealth. This emphasis on getting the **institutional** framework right is vital for ensuring long-term sustainability. It also means that small-scale, short-term conservation projects will be increasingly out of favour with donors and governments. There is a recognition that in order to future-proof conservation strategies in the face of increasing pressures, such as climate change, growing and shifting human populations and invasive species, conservation needs to be integrated within a country's **landscape planning** (Box 7.9). This is surely right—but we mustn't forget that small-scale projects, with passionate leaders and running on a tiny budget, have always been an incredibly effective way to galvanise action to conserve species or areas that people care about. The challenge is to find a way to support these interventions within a coherent broader framework.

There is one area which we feel is widely neglected in conservation, development and natural resource management, but is particularly crucial for the conservation of exploited species—gaining an understanding of an individual's incentives to act in a particular way towards natural resources. In natural resource management, the emphasis is almost entirely on the manager's best policy for optimising management outcomes. In conservation and development, there is a tendency to talk about 'the community' as if it is a homogeneous group of people. However, the person who decides to harvest, trade or use a particular plant or animal is an individual whose incentives are shaped by the social and physical world around them. Similarly, the individual who decides actively to protect an area or species has had their incentives shaped by their family, their education, or by their contact with inspirational conservationists. Thinking about individual incentives, and especially their heterogeneity, leads us to devise ways to influence these incentives, and understanding why our actions may have unintended consequences. An **incentives-based approach** is the fundamental building block for successful conservation.

In this book we have emphasised the importance of developing a **quantitative approach** to conservation science, through the use of statistically robust methods of data collection and analysis, mathematical models and decision analysis. We have also emphasised the importance of considering both the **social and biological** aspects of conservation, which is particularly important when human use is the main driver of population decline. By taking a quantitative approach, we can produce robust, defensible and generalisable research that feeds into a transparent management process. A quantitative approach does not require a strong mathematical background—it is more a state of mind. We also place emphasis on the importance of setting clear goals, **monitoring** our actions, evaluating the outcomes and learning from them. Although we can find no examples in the conservation literature in which people have used all these tools from start to finish, perhaps this is not unexpected; there are excellent examples of each component available, which we have highlighted throughout the book.

Conservation modellers tend to work in isolation—they rarely produce models as a response to requests for advice from users, follow their recommendations to implementation, learn from the outcomes, revise their models accordingly and start the process again. Conservation managers, almost without exception, do not first use models to understand the consequences of their assumptions about the system, and to help them to avoid pitfalls. This is a serious issue which leads to a trial-and-error approach to conservation action, and so to a longer learning period than would otherwise be necessary. By giving a comprehensive, though necessarily superficial, overview of the quantitative methods which are useful for conserving and managing a harvested resource, we hope to have contributed in some way to improving the sustainability of human interactions with our environment.

7.8 Resources

7.8.1 Websites

Participatory Monitoring:

Participatory assessment, monitoring and evaluation of biodiversity (online workshop): http://www.etfrn.org/etfrn/workshop/biodiversity/

Participatory monitoring case studies: http://www.monitoringmatters.org/

Participatory Planning, Monitoring and Evaluation (planning and managing development projects): http://portals.wdi.wur.nl/ppme/

Putting Fisher's Knowledge to Work. Conference, 27–31 Aug. 2001, Fisheries Centre, UBC, Vancouver, Canada. www.fisheries.ubc.ca/publications/reports/ 11-1/11-1B.pdf

Co-management:

Collective action and property rights: http://www.capri.cgiar.org/capri.htm
DFID Fisheries Management Science Programme: http://www.fmsp.org.uk/ Home.htm

Evaluating conservation success:

Conservation Evidence (database of conservation successes and failures): http://www.conserva-tionevidence.com/
Foundations of success (resource for monitoring of conservation effectiveness): http://www.fos-online.org/Site_Home.cfm
International Fund for Agricultural Development (Project monitoring & evaluation guide) http://www.ifad.org/evaluation/guide/index.htm

Other:

TRAFFIC (international wildlife trade monitor) http://www.traffic.org/
Precautionary principle project (how to apply the precautionary principle in conservation): http://www.pprinciple.net/
Precautionary approach to fisheries. Part 2: scientific papers. Prepared for the Technical Consultation on the Precautionary Approach to Capture Fisheries (Including Species Introductions). Lysekil, Sweden, 6–13 June 1995. (A scientific meeting organized by the Government of Sweden in cooperation with FAO). FAO Fisheries Technical Paper. No. 350, Part 2. Rome, FAO. 1996. http://www.fao.org/DOCREP/003/W1238E/W1238E00.htm#TOC
Adaptive management resources: http://www.adaptivemanagement.net/resources.php

7.8.2 Textbooks

Burgman, M. (2005). *Risks and Decisions for Conservation Management*. Cambridge University Press, UK. A good quantitative and theoretical treatment of the concept of risk, with later chapters explaining the principles of decision analysis.

Hilborn, R., and Mangel, M. (1997). *The Ecological Detective: Confronting Models with Data*. Princeton University Press, Princeton, N.J. Excellent book on how to analyse real world biological data.

Margoluis, R., and Salafsky, N. (1998). *Measures of Success: Designing, Managing, and Monitoring Conservation and Development Projects*. Island Press, Washington DC. A readable guide to set-ting up and implementing conservation projects, focussing particularly on developing mea-surable objectives and a sound management plan. Assumes no knowledge, and does not go into technical detail.

McShane, T.O., and Wells, M.P. (2004). *Getting Biodiversity Projects to Work: Towards More Effective Conservation and Development*. Columbia University Press, New York. An edited volume giving case studies of integrated conservation and development projects and drawing lessons from their outcomes.

Oates, J.F. (1999). *Myth and Reality in the Rain Forest: How Conservation Strategies Are Failing in West Africa*. California University Press, Berkeley. A chance to get a personal perspective on actual conservation project implementation from an experienced conservationist. Biassed, and unfashionable in many of his views, but that is exactly why it's a valuable read.

Ostrom, E. (1990). *Governing the commons: The Evolution of Institutions for Collective Action*. Cambridge University Press, UK. A classic text outlining the institutions that are needed for effective governance of common property resources.

Pilz, D., Ballard, H.L., and Jones, E.T. (2005). *Broadening participation in biological monitoring: Guidelines for Biologists and Managers*. Institute for Culture and Ecology.

http://www.ifcae.org/projects/ncssf3/guidelines.htm. Very comprehensive online textbook, with a very user-friendly annotated bibliography. Focuses on the practicalities rather than the scientific issues.

Pomeroy, R., and Rivera-Guieb, R. (2006). *Fishery Co-Management: A Practical Handbook.* International Development Research Centre and CABI Publishing. *http://www.idrc.ca/openebooks/184-1/* Freely available online textbook. Much more than the title suggests: lots of practical and transferable advice about how to interact with stakeholders.

Bibliography

Abbott, J.I.O., and Mace, R. (1999). Managing protected woodlands: Fuelwood collection and law enforcement in Lake Malawi National Park. *Conservation Biology*, 13, 418–421.

Adams, W.M. (2004). Against Extinction: The story of conservation. Earthscan, London.

Alcala, A.C. (1998). Community-based coastal resources management in the Philippines: A case study. *Ocean and Coastal Management*, 38, 179–186.

Alvard, M.S. (1993). Testing the ecologically noble savage hypothesis—Interspecific prey choice by Piro hunters of Amazonian Peru. *Human Ecology*, 21, 355–387.

Amstrup, S., MacDonald, L., and Manly, B. (2006). *Handbook of Capture-Recapture Analysis*. Princeton University Press, Princeton, NJ.

Anderies, J.M., Janssen, M.A., and Ostrom, E. (2004). A framework to analyze the robustness of social-ecological systems from an institutional perspective. *Ecology & Society*, 9, 192–208. http://www.ecologyandsociety.org/vol9/iss1/art18/print.pdf

Anderson, J., Rowcliffe, J.M., and Cowlishaw, G. (2007). The Angola black-and-white colobus (*Colobus angolensis palliatus*) in Kenya: Historical range contraction and current conservation status. *American Journal of Primatology*, 69, 1–17.

Andrianandrasana, H.T., Randriamahefasoa, J., Durbin, J., Lewis, R.E., and Ratsimbazafy, J.H. (2005). Participatory ecological monitoring of the Alaotra wetland in Madagascar. *Biodiversity & Conservation*, 14, 2757–2774. www.monitoringmatters.org.

Arthur, R.I., and Howard, C. (2005). *Co-Management: A Synthesis of Lessons Learnt from the DFID Fisheries Comanagement Science Programme*. MRAG Ltd, London. http://www.fmsp. org.uk/Documents/keylessons/R8470_Ann3-2.pdf

Ashworth, J.S., and Ormond, R.F.G. (2005). Effects of fishing pressure and trophic group on abundance and spillover across boundaries of a no-take zone. *Biological Conservation*, 121, 333–344.

Aswani, S., and Hamilton, R.J. (2004). Integrating indigenous ecological knowledge and customary sea tenure with marine and social science for conservation of bumphead parrotfish (*Bolbometopon muricatum*) in the Roviana Lagoon, Solomon Islands. *Environmental Conservation*, 31, 69–83. http://www.anth.ucsb.edu/faculty/aswani/articles/IEK-CSTpaper.pdf

Aswani, S., and Lauer, M. (2006). Incorporating fishers' local knowledge and behaviour into Geographical Information Systems (GIS) for designing marine protected areas in Oceania. *Human Organization*, 65(1), 80–101. http://www.anth.ucsb.edu/faculty/aswani/articles/ Aswani-Lauer-HO-2006.pdf

Auzel, P. and Wilkie, D. (2000). Wildlife use in northern Congo: Hunting in a commercial logging concession. In Robinson, J.R. and Bennett, E.L., eds. *Hunting for Sustainability in Tropical Forests*, pp. 413–426. Columbia University Press, New York, NY.

Ayres, J.M., de Magalhaes Lima, D., de Souza Martins, E., and Barreiros, J.L.K. (1991). On the track of the road: Changes in subsistence hunting in a Brazilian Amazonian village. In Robinson, J.R. and Redford, K., eds. *Neotropical Wildlife Use and Conservation*. Chicago University Press Chicago, IL, pp. 82–92.

Baker, J. (2004). Evaluating Conservation Policy: Integrated Conservation and Development in Bwindi Impenetrable National Park Uganda. PhD thesis, University of Kent, UK.

Balmford, A., and Whitten, T. (2003). Who should pay for tropical conservation, and how could the costs be met? *Oryx*, 37, 238–250.

Bamberg, S. (2003). How does environmental concern influence specific environmentally-related behaviours? A new answer to an old question. *Journal of Environmental Psychology*, 23, 21–32.

Bann, C. (1998). *The Economic Valuation of Mangroves: A Handbook for Researchers*. International Development Research Centre, Ottawa, Canada. http://web.idrc.ca/uploads/user-S/10305674900acf30c.html

Barlow, E.D., and Bakke, A.N. (n.d.). *Managing Alaska's Halibut: Observations from the Fishery*. Environmental Defense Fund. http://www.environmentaldefense.org/documents/489_halibut.PDF

Barrett, C.B., and Arcese, P. (1998). Wildlife harvest in integrated conservation and development projects: Linking harvest to household demand, agricultural production, and environmental shocks in the Serengeti. *Land Economics*, 74, 449–465.

BCTF (2005). *Bushmeat Education Resource Guide*. Bushmeat Crisis Task Force, Washington, DC. http://www.bushmeat.org/cd/berg.html

BCTF (n.d.). *Goals*. Bushmeat Crisis Task Force, Washington, DC. http://www.bushmeat.org/goals.html

Beger, M., Harborne, A.R., Dacles, T.P., Solandt, J.L., and Ledesma, G.L. (2004). A framework of lessons learned from community-based marine reserves and its effectiveness in guiding a new coastal management initiative in the Philippines. *Environmental Management*, 34, 786–801.

Begg, D., Fischer, S., and Dornbusch, R. (2005). *Economics*, Eighth Edition. McGraw-Hill Columbus, OH.

Begon, M., Townsend, C., and Harper, J.L. (2005). *Ecology*, Fourth Edition. Blackwell Science, Oxford.

Béné, C. (2003). When fishery rhymes with poverty: A first step beyond the old paradigm on poverty in small-scale fisheries. *World Development*, 31, 949–975.

Bennett, E.L., Blencowe, E., Brandon, K., Brown, D., Burn, R.W., Cowlishaw, G.C., *et al.* (2007). Hunting for Consensus: A statement on reconciling bushmeat harvest, conservation and development policy in west and central Africa. *Conservation Biology*, 21, 884–887.

Benton, T.G., and Grant, A. (1999). Elasticity analysis as an important tool in evolutionary and population ecology. *Trends in Ecology and Evolution*, 14, 467–471.

Berkes, F., Davidson-Hunt, I., and Davidson-Hunt, K. (1998). Diversity of common property resource use and diversity of social interests in the western Indian Himalaya. *Mountain Research and Development*, 18, 19–33. http://home.cc.umanitoba.ca/~berkes/berkes_1998_3.pdf

Bibby, C.J., Burgess, N.D., and Hill, D.A. (1992). *Bird Census Techniques*. Academic Press: London.

Binmore, K. (2007). *Playing for Real: Game Theory*. Oxford University Press: Oxford.

Bishir, J., and Lancia, R.A. (1996). On catch-effort methods of estimating animal abundance. *Biometrics*, 52, 1457–1466.

Blundell, A.G. (2004). A review of the CITES listing of big-leaf mahogany. *Oryx*, 38, 84–90.

Bodkin, J.L., Ames, J.A., Jameson, R.J., Johnson, A.M., and Matson, G.M. (1997). Estimating age of sea otters with cementum layers in the first premolar. *Journal of Wildlife Management*, 61, 967–973.

Borchers, D.L., Buckland, S.T., and Zucchini, W. (2002). *Estimating Animal Abundance: Closed Populations*. Springer-Verlag: London.

Borchers, D.L., Buckland, S.T., Goedhart, P.W., Clarke, E.D., and Hedley, S.L. (1998). Horvitz-Thompson estimators for double-platform line transect surveys. *Biometrics*, 54, 1221–1237.

Born Free Foundation (2002). *Stop the Slaughter—Ban the Bloody Ivory Trade*. Press Release. http://www.engology.com/articleivory.htm

Borrini-Feyerabend, G., Kothari, A., and Oviedo, G. (2004). *Protected Areas: Towards Equity and Enhanced Conservation*. Guidance on policy and practice for Co-managed Protected Areas and Community Conserved Areas. Best Practice Protected Area Guidelines Series No. 11. World Commission on Protected Areas (WCPA), IUCN—The World Conservation Union 2004. ISBN: 2-8317-0675-0. http://www.iucn.org/bookstore/HTML-books/ BP11-indigenous_and_local_communities/cover.html

Bousquet, F., Le Page, C., Bakam, I., and Takforyan, A. (2001). Multiagent simulations of hunting wild meat in a village in eastern Cameroon. *Ecological Modelling*, 138, 331–346.

Boyce, M.S. (2006). Scale for resource selection functions. *Diversity and Distributions*, 12, 269–276.

Boyer, D.C., Kirchner, C.H., McAllister, M.K., Staby, A., and Staalesen, B.I. (2001). The orange roughy fishery of Namibia: Lessons to be learned about managing a developing fishery. *South African Journal of Marine Science*, 23, 205–221.

Bradshaw, C.J.A., Barker, R.J., and Davis, L.S. (2000). Modeling tag loss in New Zealand fur seal pups. *Journal of Agricultural Biological and Environmental Statistics*, 5, 475–485.

Branch, T.A., Hilborn, R., Haynie, A.C., Fay, G., Flynn, L., Griffiths, J., *et al.* (2006). Fleet dynamics and fishermen behavior: Lessons for fisheries managers. *Canadian Journal of Fisheries and Aquatic Sciences*, 63, 1647–1668.

Brashares, J.S., Arcese, P., Sam, M.K., Coppolillo, P.B., Sinclair, A.R.E., and Balmford, A. (2004). Bushmeat hunting, wildlife declines, and fish supply in West Africa. *Science*, 306, 1180–1183.

Brickle, N.W. (2002). Habitat use, predicted distribution and conservation of green peafowl (*Pavo muticus*) in Dak Lak Province, Vietnam. *Biological Conservation*, 105, 189–197.

Brook, B.W., and Bradshaw, C.J.A. (2006). Strength of evidence for density dependence in abundance time series of 1198 species. *Ecology*, 87, 1445–1451.

Brook, B.W., O'Grady, J.J., Chapman, A.P., Burgman, M.A., Akcakaya, H.R., and Frankham, R. (2000). Predictive accuracy of population viability analysis in conservation biology. *Nature*, 404, 385–387.

Brooke, A.P. (2001). Population status and behaviours of the Samoan flying fox (*Pteropus samoensis*) on Tutuila Island, American Samoa. *Journal of Zoology*, 254, 309–319.

Brooke, A.P., and Tschapka, M. (2002). Threats from overhunting to the flying fox, *Pteropus tonganus* (Chiroptera: Pteropodidae) on Niue Island, South Pacific Ocean. *Biological Conservation*, 103, 343–348.

Brooks, J.S., Franzen, M.A., Holmes, C.M., Grote, M.N., and Mulder, M.B. (2006, October). Testing hypotheses for the success of different conservation strategies. *Conservation Biology*, 20(5), 1528–1538.

Buckland, S.T., Anderson, D.R., Burnham, K.P., Laake, J.L., Borchers, D.L., and Thomas, L. (2004). *Introduction to Distance Sampling: Estimating Abundance of Biological Populations*. Oxford University Press: Oxford.

Buckland, S.T., Summers, R.W., Borchers, D.L., and Thomas, L. (2006). Point transect sampling with traps or lures. *Journal of Applied Ecology*, 43, 377–384.

Buckmeier, D.L., and Howells, R.G. (2003). Validation of otoliths for estimating ages of largemouth bass to 16 years. *North American Journal of Fisheries Management*, 23, 590–593.

Burnham, K.P., and Anderson, D.R. (2002). *Model Selection and Multi-Model Inference: A Practical Information-Theoretic Approach*. Springer-Verlag: New York, NY.

Burton, J.A., Hedges, S., and Mustari, A.H. (2005). The taxonomic status, distribution and conservation of the lowland anoa *Bubalus depressicornis* and mountain anoa *Bubalus quarlesi*. *Mammal Review*, 35, 25–50.

Cain, J.D., Jinapala, K., Makin, I.W., Somaratna, P.G., Ariyaratna, B.R., and Perera, L.R. (2003). Participatory decision support for agricultural management. A case study from Sri Lanka. *Agricultural Systems*, 76, 457–482.

Caldecott, J. (1988). *Hunting and Wildlife Management in Sarawak*. IUCN: Cambridge, UK.

Calvert, A.M., and Gauthier, G. (2005). Effects of exceptional conservation measures on survival and seasonal hunting mortality in greater snow geese. *Journal of Applied Ecology*, 42, 442–452.

Cardenas, J.C. (2004). Norms from outside and from inside: An experimental analysis on the governance of local ecosystems. *Forest Policy and Economics*, 6, 229–241.

Carroll, C., Zielinski, W.J., and Noss, R.F. (1999). Using presence-absence data to build and test spatial habitat models for the fisher in the Klamath region, U.S.A. *Conservation Biology*, 13, 1344–1359.

Carvalho, S., and White, H. (1997). *Combining the Quantitative and Qualitative Approaches to Poverty Management and Analysis: The Practice and the Potential*. World Bank: Washington, D.C. ISBN: 0-8213-3955-9.

Caswell, H. (2001). *Matrix Population Models: Construction Analysis and Interpretation*, 2nd edn. Sinauer: Sunderland, MA.

Caughley, G., and Gunn, A. (1996). *Conservation Biology in Theory and Practice*. Blackwell Science: Oxford, UK.

CEBC (2006). *Guidelines for Systematic Review in Conservation and Environmental Management*. Centre for Evidence-Based Conservation, University of Birmingham, UK. http://www.cebc.bham.ac.uk/Documents/CEBC%20Systematic%20Review%20Guidelines%20Version%202.0.pdf

Chaloupka, M. (2002). Stochastic simulation modelling of southern Great Barrier Reef green turtle population dynamics. *Ecological Modelling*, 148, 79–109.

Chambers, R. (1992). *Rural Appraisal: Rapid, Relaxed and Participatory*. IDS Discussion papers 311, Institute for Development Studies, UK. http://www.ids.ac.uk/ids/bookshop/dp/dp311.pdf

Chambers, R. (2003). *Notes for Participants in PRA-PLA Familiarisation Workshops in 2003*. Participation Resource Centre, Institute for Development Studies, UK. http://www.ids.ac.uk/ids/particip/research/pra/pranotes03.pdf

Charles, A.T. (2002). Use rights and responsible fisheries: Limiting access and harvesting through rights-based management. In Cochrane, K., ed. *A Fishery Manager's Guidebook: Management Measures and Their Application*, Chapter 6, pp. 131–158. Food and Agriculture Organisation Technical Paper number 424. FAO, Rome. ftp://ftp.fao.org/docrep/fao/004/y3427e/y3427e00.pdf

Chhatre, A., and Saberwal, V. (2005). Political incentives for biodiversity conservation. *Conservation Biology*, 19, 310–317.

CIC (2002). *Sustainable Hunting: An Instrument for Species Protection and to Fight Poverty*. International Council for Game and Wildlife Conservation. http://www.cic-wildlife.org/index.php?id=35

Cinner, J.E., and McClanahan, T.R. (2006). Socioeconomic factors that lead to overfishing in small-scale coral reef fisheries of Papua New Guinea. *Environmental Conservation*, 33, 73–80.

CITES (n.d.). *Monitoring the Illegal Killing of Elephants* (MIKE). Convention on Illegal Trade in Endangered Species of Wild Fauna and Flora Secretariat. http://www.cites.org/eng/prog/MIKE/intro/index.shtml

Clark, C.W. (1990). *Mathematical Bioeconomics*. John Wiley & Son, New York, NY.

Clark, C.W., and Kirkwood, G.P. (1986). On uncertain renewable resource stocks: Optimal harvest policies and the value of stock surveys. *Journal of Environmental Economics and Management*, 13, 235–244.

Clayton, L.M., Keeling, M., and Milner-Gulland, E.J. (1997). Bringing home the bacon: A spatial model of wild pig harvesting in Sulawesi, Indonesia. *Ecological Applications*, 7, 642–652.

Clayton, L.M., Milner-Gulland, E.J., Sinaga, D.W., and Mustari, A.H. (2000). Effects of a proposed ex situ conservation programme on in situ conservation of the babirusa. *Conservation Biology*, 14, 382–385.

Cochrane, K., ed. (2002). *A Fishery Manager's Guidebook: Management Measures and Their Application*. Food and Agriculture Organisation Technical Paper number 424. FAO, Rome. ftp://ftp.fao.org/docrep/fao/004/y3427e/y3427e00.pdf

Coffey, A., Holbrook, B., and Atkinson, P. (1996). *Qualitative Data Analysis: Technologies and Representations*. Sociological Research Online 1. http://www.socresonline.org.uk/socresonline/1/1/4.html

Cohen, J. (1992). *Statistical Power Analysis for the Behavioural Sciences*, Second Edition. Erlbaum: Hillsdale, NJ. ISBN: 0-8058-0283-5.

Cole, L.C. (1954). The population consequences of life history phenomena. *Quarterly Review of Biology*, 29, 103–137.

Coltman, D.W., O'Donoghue, P., Jorgenson, J.T., Hogg, J.T., Strobeck, C., and Festa-Bianchet, M. (2003). Undesirable evolutionary consequences of trophy hunting. *Nature*, 426, 655–658.

Convention on Biological Diversity (1993). *Convention on Biological Diversity (with Annexes)*. Concluded at Rio de Janeiro on 5 June 1992. http://www.biodiv.org/convention/articles.shtml?a=cbd-02

Cooch, E., and White, G.C., eds. (2006). *Program MARK: A Gentle Introduction*. http://www.phidot.org/software/mark/docs/book/

Costello, C.M., Inman, K.H., Jones, D.E., Inman, R.M., Thompson, B.C., and Quigley, H.B. (2004). Reliability of the cementum annuli technique for estimating age of black bears in New Mexico. *Wildlife Society Bulletin*, 32, 169–176.

Coulson, T., Mace, G.M., Hudson, E., and Possingham, H. (2001). The use and abuse of population viability analysis. *Trends in Ecology and Evolution*, 16, 219–221.

Courchamp, F., Angulo, E., Rivalan, P., Hall, R.J., Signoret, L., Bull, L., *et al.* (2006, December). Rarity value and species extinction: The anthropogenic Allee effect. *Public Library of Science Biology*, 4(12), e415. http://www.pubmedcentral.nih.gov/articlerender.fcgi?artid=1661683

Cowling, R.J., and Pressey, R.L. (2003). Introduction to systematic conservation planning in the Cape Floristic Region. *Biological Conservation*, 112, 1–13.

Cowlishaw, G.C., Mendelson, S., and Rowcliffe, J.M. (2005a). Evidence for post-depletion sustainability in a mature bushmeat market. *Journal of Applied Ecology*, 42, 460–468.

Cowlishaw, G.C., Mendelson, S., and Rowcliffe, J.M. (2005b). Structure and operation of a bushmeat commodity chain in southwestern Ghana. *Conservation Biology*, 19, 139–149.

Crawley, M.J. (2005). *Statistics: An Introduction Using R*. John Wiley: Chichester.

Crawley, M.J. (2007). *The R Book*. John Wiley and Sons: Chichester.

Critical Ecosystems Partnership Fund (2005). *A Review of CEPF's Portfolio in the Cape Floristic Region*. http://www.cepf.net/ImageCache/cepf/content/pdfs/cepf_2ecapefloristicregion_2eoverview_5f3_2e05_2epdf/v1/cepf.capefloristicregion.overview_5f3.05.pdf

Crookes, D.J., Ankudey, N., and Milner-Gulland, E.J. (2006). The value of a long-term bushmeat market dataset as an indicator of system dynamics. *Environmental Conservation*, 32, 333–339.

Crouse, D.T., Crowder, L.B., and Caswell, H. (1987). A stage-based population model for loggerhead sea-turtles and implications for conservation. *Ecology*, 68, 1412–1423.

Damania, R., Milner-Gulland, E.J., and Crookes, D.J. (2005). A bioeconomic model of bushmeat hunting. *Proceedings of the Royal Society B*, 272, 259–266.

Damania, R., Stringer, R., Karanth, K.U., and Stith, B. (2003). The economics of protecting tiger populations: Linking household behavior to poaching and prey depletion. *Land Economics*, 79, 198–216.

Danielsen, F., Burgess, N.D., and Balmford, A. (2005). Monitoring matters: Examining the potential of locally-based approaches. *Biodiversity & Conservation*, 14, 2507–2542. www.monitoringmatters.org

Dauer, J.T., Mortensen, D.A., and Vangessel, M.J. (2007). Temporal and spatial dynamics of long-distance Conyza canadensis seed dispersal. *Journal of Applied Ecology*, 44, 105–114.

Davies, G., and Brown, D. (in press). The consumptive use of wildlife in the tropics: Challenges for the 21st century. Blackwell Science: Oxford.

de Lopez, T. T. (2003). Economics and stakeholders of Ream National Park, Cambodia. *Ecological Economics*, 46, 269–282.

de Merode, E., and Cowlishaw, G. (2006). Species protection, the changing informal economy, and the politics of access to the bushmeat trade in the Democratic Republic of Congo. *Conservation Biology*, 20, 1262–1271.

de Merode, E., Homewood, K., and Cowlishaw, G. (2004). The value of bushmeat and other wild foods to rural households living in extreme poverty in the Democratic Republic of Congo. *Biological Conservation*, 118, 573–581.

de Vaus, D.A. (2002). *Surveys in Social Research*, Fifth Edition. Allen & Unwin: Crow's Nest, Australia.

DFID (2001). *Sustainable Livelihoods Guidance Sheets*. Department for International Development, London, UK. http://www.livelihoods.org/info/info_guidancesheets.html

Dorazio, R.M., and F. A Johnson. (2003). Bayesian inference and decision theory—A framework for decision making in natural resource management. *Ecological Applications*, 13, 556–563. http://cars.er.usgs.gov/BayesExampleRev.pdf

Drew, J.A. (2005). Use of traditional ecological knowledge in marine conservation. *Conservation Biology*, 19, 1286–1293.

Dublin, H.T., Milliken, T., and Barnes, R.F.W. (1994). *Four Years After the CITES Ban: Illegal Killing of Elephants, Ivory Trade and Stockpiles*. IUCN/SSC African elephant specialist group, Nairobi.

Durbin, J. (2003). *Conservation Contracts: Direct Incentives to Communities for Biodiversity Conservation in Madagascar*. Presentation at the World Parks Congress, S. 3 Workshop 9: Conservation Incentive Agreements, 12th September 2003. http://www.conservationfinance. org/WPC/WPC_documents/Apps_09_%20Durbin_ppt_v1.pdf

East, T., Kumpel, N., Milner-Gulland, E.J., and Rowcliffe, J.M. (2005). Determinants of urban bushmeat consumption in Rio Muni, Equatorial Guinea. *Biological Conservation*, 126, 206–215.

Eberhardt, L.L. (1969). Population analysis. In Giles, R.H. Jr, ed. *Wildlife Management Techniques*. The Wildlife Society: Washington, DC. pp. 457–95.

Eberhardt, L.L. (2002). A paradigm for population analysis of long-lived vertebrates. *Ecology*, 83, 2841–2854.

Ecott, T. (2006, March 18). Easy on Vamizi. *The Guardian Newspaper*. http://travel.guardian. co.uk/article/2006/mar/18/mozambique.ecotourism.hotels

Edgar, G.J., Barrett, N.S., and Morton, A.J. (2004). Biases associated with the use of underwater visual census techniques to quantify the density and size-structure of fish populations. *Journal of Experimental Marine Biology and Ecology*, 308, 269–290.

Efford, M.G. (2004). Density estimation in live-trapping studies. *Oikos*, 106, 598–610.

EIA-WPSI (2006). *Skinning the Cat: Crime and Politics of the Big Cat Skin Trade*. Environmental Investigation Agency & Wildlife Protection Society of India. http://www.wpsi-india.org/images/EIA-WPSI_Skinning_The_Cat.pdf

Elkan, P., and Elkan, S. (2002, November). Engaging the private sector in the bushmeat crisis. *American Zoo and Aquarium Association Communiqué*, 40–43. http://www.aza.org/ Publications/2002/11/Nov2002BushmeatPrivateSector.pdf

Elmes, G.W., and Thomas, J.A. (1992). Complexity of species conservation in managed habitats: Interactions between *Maculinea* butterflies and their ant hosts. *Biodiversity and Conservation*, 1, 155–169.

Elton, C.S., and Nicholson, A.J. (1942). The ten year cycle in the numbers of lynx in Canada. *Journal of Animal Ecology*, 11, 215–244.

English, S., Wilkinson, C., and Baker, V. (1997). *Survey Manual for Tropical Marine Resources*, Second Edition. Australian Institute of Marine Sciences: Townsville.

European Commission (2003). *CAP Reform—A Long-Term Perspective for Sustainable Agriculture*. http://europa.eu.int/comm/agriculture/capreform/index_en.htm

Ezard, T.H.C., Becker, P.H., and Coulson, T. (2006). The contributions of age and sex to variation in common term population growth rate. *Journal of Animal Ecology*, 75, 1379–1386.

Fa, J.E., Currie, D., and Meeuwig, J. (2003). Bushmeat and food security in the Congo Basin: Linkages between wildlife and people's future. *Environmental Conservation*, 30, 71–78.

Fa, J.E., Juste, J., Burn, R.W., and Broad, G. (2002). Bushmeat consumption and preferences of two ethnic groups in Bioko Island, West Africa. *Human Ecology*, 30, 397–416.

FAO (1996). *Towards Sustainable Food Security: Women and Sustainable Food Security*. SD Dimensions, Food and Agriculture Organisation, Rome. http://www.fao.org/sd/fsdirect/fbdirect/FSpecial.htm

FAO (2001). *Global Forest Resources Assessment 2000*. Food and Agriculture Organisation, Rome, Italy. http://www.fao.org/forestry/site/fra2000report/en/

Feer, F. (1991). The potential for sustainable hunting & rearing of game in tropical forests. In Hladik, C.M., Hladik, A., Linares, O.F., Pagezy, H., Semple, A., and Hadley, M. eds. *Tropical Forests, People & Food: Biocultural Interactions and Applications to Development*, Volume 13. UNESCO MAB Series UNESCO: Paris, pp. 691–708.

Ferraro, P.J. (2001). Global habitat protection: Limitations of development interventions and a role for conservation performance payments. *Conservation Biology*, 15, 990–1000.

Ferraro, P.J., and Kiss, A. (2002). Direct payments to conserve biodiversity. *Science*, 298, 1718–1719.

Fieberg, J., and Jenkins, K.J. (2005). Assessing uncertainty in ecological systems using global sensitivity analyses: A case example of simulated wolf reintroduction effects on elk. *Ecological Modelling*, 187, 259–280.

Field, B., and Field, M. (1997). *Environmental Economics: An Introduction*, Third Edition. McGraw-Hill: New York, NY.

Forget, P.-M., and Jansen, P.A. (2007). Hunting increases dispersal limitation in the tree *Carapa procera*, a nontimber forest product. *Conservation Biology*, 21, 106–113.

Fowler, C.W. (1981). Density dependence as related to life history strategy. *Ecology*, 62, 602–610.

Freckleton, R.P., Watkinson, A.R., Green, R.E., and Sutherland, W.J. (2006). Census error and the detection of density dependence. *Journal of Animal Ecology*, 75, 837–851.

Frederiksen, M., and Bregnballe, T. (2000). Evidence for density-dependent survival in adult cormorants from a combined analysis of recoveries and resightings. *Journal of Animal Ecology*, 69, 737–752.

Freehling, J., and Marks, S. (1998). *A century of change in the Central Luangwa Valley of Zambia*. In Milner-Gulland, E.J. and Mace, R. eds. *Conservation of Biological Resources*. Blackwell Science: Oxford, pp. 261–278.

Freeman, M.M.R. (1994). *Gallup on Public Attitudes to Whales and Whaling*. The High North Alliance, Canada. http://www.highnorth.no/Library/Opinion/ga-on-pu.htm.

Fryxell, J., Falls, J.B., Falls, E.A., Brooks, R.J., Dix, L., and Strickland, M. (2001). Harvest dynamics of mustelid carnivores in Ontario, Canada. *Wildlife Biology*, 7, 151–159.

Fuller, T.K., York, E.C., Powell, S.M., Decker, T.A., and DeGraaf, R.M. (2001). An evaluation of territory mapping to estimate fisher density. *Canadian Journal of Zoology-Revue Canadienne De Zoologie*, 79, 1691–1696.

Gaidet-Drapier, N., Fritz, H., Bourgarel, M., Renaud, P-C., Poilecot, P., Chardonnet, P., *et al.* (2006). Cost and efficiency of large mammal census techniques: Comparison of methods for a participatory approach in a communal area, Zimbabwe. *Biodiversity & Conservation*, 15, 735–754.

Gaillard, J.-M., Festa-Bianchet, M., and Yoccoz, N.G. (1998). Population dynamics of large herbivores: Variable recruitment with constant adult survival. *Trends in Ecology and Evolution*, 13, 58–63.

Gaillard, J.-M., and Yoccoz, N.G. (2003). Temporal variation in survival of mammals: A case of environmental canalization? *Ecology*, 84, 3294–3306.

Garaway, C., and Arthur, R. (2004). *Adaptive Learning: A Practical Framework for the Implementation of Adaptive Co-Management—Lessons from Selected Experiences in South and South-East Asia*. MRAG Ltd., London. http://www.onefish.org/servlet/BinaryDownloaderServlet? filename=1115370074421_Adaptive_learning_guidelines.pdf&refID=248931

Gell, F.R., and Roberts, C.M. (2003). *The Fishery Effects of Marine Reserves and Fishery Closures*. World Wildlife Fund: Washington, DC. http://www.worldwildlife.org/oceans/pdfs/ fishery_effects.pdf

Gerber, L.R., Botsford, L., Hastings, A., Possingham, H., Gaines, S., Palumbi, S., *et al.* (2003). Population models for reserve design: A retrospective and prospective synthesis. *Ecological Applications* (Special Issue on Marine Reserves) 13, S47-S64. http://www.public.asu. edu/~lrgerbe/publications.htm

Getz, W.M., and Haight, R.G. (1989). *Population harvesting: Demographic models of fish, forest, and animal resources*. Princeton University Press: Princeton, NJ.

Gezelius, S.S. (2004). Food, money and morals: Compliance among natural resource harvesters. *Human Ecology*, 32, 615–634.

Ghazoul, J., and McAllister, M. (2003). Communicating complexity and uncertainty in decision making contexts: Bayesian approaches to forest research. *International Forestry Review*, 5, 9–19.

Gilbert, G., McGregor, P.K., and Tyler, G. (1994). Vocal individuality as a census tool— Practical considerations illustrated by a study of two rare species. *Journal of Field Ornithology*, 65, 335–348.

Gilmour, D., Malla, Y., and Nurse, N. (2004). *Linkages Between Community Forestry and Poverty*. Position paper, RECOFTC, Thailand. http://www.iucn.org/themes/ceesp/publications/ SL/Community%20forestry%20and%20poverty%20RECOFTC.pdf

Ginsberg, J.R., and Milner-Gulland, E.J. (1994). Sex-biassed harvesting and population dynamics: Implications for conservation and sustainable use. *Conservation Biology*, 8, 157–166. www.iccs.org.uk

Gipson, P.S., Ballard, W.B., Nowak, R.M., and Mech, L.D. (2000). Accuracy and precision of estimating age of gray wolves by tooth wear. *Journal of Wildlife Management*, 64, 752–758.

Gole, C., Burton, M., Williams, K.J., Clayton, H., Faith, D.P., White, B., *et al.* (2005). *Auction for Landscape Recovery: Final Report*. WWF-Australia, Sydney. http://www.wwf.org.au/ publications/AuctionForLandscapeRecovery/

Gomez-Aparicio, L., Gomez, J.M., and Zamora, R. (2007). Spatiotemporal patterns of seed dispersal in a wind-dispersed Mediterranean tree (*Acer opalus* subsp *granatense*): Implications for regeneration. *Ecography*, 30, 13–22.

Gray, M., and Kalpers, J. (2005). Ranger based monitoring in the Virunga-Bwindi region of East-Central Africa: A simple data collection tool for park management. *Biodiversity & Conservation*, 14, 2723–2741. www.monitoringmatters.org

Grimble, R. (1998). *Stakeholder Methodologies in Natural Resource Management.* Socioeconomic Methodologies: Best Practice Guidelines. Natural Resources Institute: Chatham, UK. http://www2.dfid.gov.uk/pubs/files/BPG02.pdf

Guisan, A., and Thuiller, W. (2005). Predicting species distribution: Offering more than simple habitat models. *Ecology Letters*, 8, 993–1009.

Gunn, A. (1998). Caribou and muskox harvesting in the Northwest Territories. In Milner-Gulland, E.J., and Mace, R., eds. *Conservation of Biological Resources*, pp. 314–330. Blackwell Science: Oxford.

Gurevitch, J., Curtis, P.S., and Jones, M.H. (2001). Meta-analysis in ecology. *Advances in Ecological Research*, 32, 199–247.

Haddad, L., Hoddinott, J., and Alderman, H. (1997). *Intrahousehold Resource Allocation in Developing Countries.* IFPRI Food Policy Statement 24. International Food Policy Research Institute, http://www.ifpri.org/pubs/fps/ps24.asp

Halls, A.S., Arthur, R., Bartley, D., Felsing, M., Grainger, R., Hartmann, W., *et al.* (2005). *Guidelines for Designing Data Collection and Sharing Systems for Co-Managed Fisheries.* Part 2: Technical guidelines. FAO Fisheries Technical Paper. No. 494/2. Rome, FAO. ftp://ftp.fao.org/docrep/fao/008/a0231e/a0231e00.pdf

Halls, A.S., Burn, R.W., and Abeyasekera, S. (2002). *Interdisciplinary Multivariate Analysis for Adaptive Co-Management DFID project r7834.* Final technical report. DFID Fisheries Management Science Programme. http://p15166578.pureserver.info/fmsp/Documents/r7834/r7834_4.pdf

Hammond, T.R. and O'Brien, C.M. (2001). An application of the Bayesian approach to stock assessment model uncertainty. *ICES Journal of Marine Science*, 58, 648–656.

Hanley, N., Shogren, J., and White, B. (2001). *Introduction to Environmental Economics.* Oxford University Press: UK.

Hardin, G. (1968). The tragedy of the commons. *Science*, 162, 1243–1248.

Harley, S.J., Myers, R.A., and Dunn, A. (2001). Is catch-per-unit-effort proportional to abundance? *Canadian Journal of Fisheries and Aquatic Sciences*, 58, 1760–1772.

Harrison, P.J., Buckland, S.T., Thomas, L., Harris, R., Pomeroy, P.P., and Harwood, J. (2006). Incorporating movement into models of grey seal population dynamics. *Journal of Animal Ecology*, 75, 634–645.

Hart, J., and Hart, T. (2003, Winter). Rules of engagement for conservation: Lessons from the Democratic Republic of Congo. *Conservation in Practice*, 4(1). http://www.conbio.org/cip/article41RUL.cfm

Haydon, D.T., Shaw, D.J., Cattadori, I.M., Hudson, P.J., and Thirgood, S.J. (2002). Analysing noisy time-series: Describing regional variation in the cyclic dynamics of red grouse. *Proceedings of the Royal Society of London Series B-Biological Sciences*, 269, 1609–1617.

Hedges, S., and Lawson, D. (2006). *Dung Survey Standards for the MIKE Programme.* CITES Monitoring the Illegal Killing of Elephants Programme, Nairobi. http://www.cites.org/common/prog/mike/survey/dung_standards.pdf

Henrich, J., McElreath, R., Barr, A., Ensminger, J., Barrett, C., Bolyanatz, A., *et al.* (2006). Costly punishment across human societies. *Science*, 312, 1767–1770.

Hestbeck, J.B., Nichols, J.D., and Malecki, R.A. (1991). Estimates of movement and site fidelity using mark-resight data of wintering Canada geese. *Ecology*, 72, 523–533.

Hewison, A.J.M., Vincent, J.P., Angibault, J.M., Delorme, D., Van Laere, G., and Gaillard, J.M. (1999). Tests of estimation of age from tooth wear on roe deer of known age: Variation within and among populations. *Canadian Journal of Zoology-Revue Canadienne De Zoologie*, 77, 58–67.

Hilborn, R., and Mangel, M. (1997). The ecological detective: Confronting models with data. *Monographs in Population Biology*, 28, Princeton University Press: Princeton, NJ.

Hilborn, R., Stokes, K., Maguire, J.J., Smith, T., Botsford, L.W., Mangel, M., *et al.* (2004). When can marine reserves improve fisheries management? *Ocean and Coastal Management,* 47, 197–205.

Hilborn, R., and Walters, C.J. (1987). A general model for simulation of stock & fleet dynamics in spatially heterogeneous fisheries. *Canadian Journal of Fisheries & Aquatic Science,* 44, 1366–1369.

Hilborn, R., and Walters, C.J. (1992). *Quantitative Fisheries Stock Assessment: Choice Dynamics and Uncertainty.* Kluwer: Dordrecht.

Hill, N.A.O. (2005). Livelihoods in an Artisanal Fishing Community and the Effect of Ecotourism. MSc thesis, Imperial College London. www.iccs.org.uk

Hirzel, A., and Guisan, A. (2002). Which is the optimal sampling strategy for habitat suitability modelling. *Ecological Modelling,* 157, 331–341.

Hobbs, N.T., and Hilborn, R. (2006). Alternatives to statistical hypothesis testing in ecology: A guide to self teaching. *Ecological Applications,* 16, 5–19.

Hockley, N.J., Jones, J.P.G., Andriahajaina, F.B., Manica, A., Rakoto, F.E., Ranambitsoa, E.H., *et al.* (2005). When should communities and conservationists monitor exploited resources? *Biodiversity and Conservation,* 14, 2795–2806. www.monitoringmatters.org.

Holmes, C.M. (2003). The influence of protected area outreach on conservation attitudes and resource use patterns: A case study from western Tanzania. *Oryx,* 37, 305–315.

Howell, D.C. (2007). *Resampling Statistics: Randomization and the Bootstrap.* Accessed April 2007, from http://www.uvm.edu/~dhowell/StatPages/Resampling/Resampling.html

Hsieh, C.H., Reiss, C.S., Hunter, J.R., Beddington, J.R., May, R.M., Sugihara, G. (2006). Fishing elevates variability in the abundance of exploited species. *Nature,* 443, 859–62.

Hudson, P.J. (1992). *Grouse in Space and Time.* Game Conservancy Trust, Fordingbridge, UK.

Hurst, A. (2004). *Barren Ground Caribou Co-Management in the Eastern Canadian Arctic: Lessons for Bushmeat Management.* ODI Wildlife Policy Briefing 5. http://www.odi.org.uk/fpeg/publications/policybriefs/wildlifepolicy/downloads/WPB5final.pdf

Hutton, J., and Leader-Williams, N. (2003). Sustainable use and incentive-driven conservation: Realigning human and conservation interests. *Oryx,* 37, 215–226.

IFAD (n.d.). *Managing for Impact in Rural Development: A Guide for Monitoring and Evaluation.* International Fund for Agricultural Development. http://www.ifad.org/evaluation/guide/index.htm

Insight (n.d.). *Desertification in Central Asia.* http://www.insightshare.org/case_study_desertification.html

IPCC (2001). *Climate Change 2001: Synthesis Report.* World Meteorological Organization, UNEP. http://www.ipcc.ch/pub/syreng.htm

Ireland, C., Malleret-King, D., and Baker, L. (2004). *Alternative Sustainable Livelihoods for Coastal Communities: A Review of Experience and Guide to Best Practice.* Report submitted to IUCN by the IDL Group. http://www.theidlgroup.com/downloads/Sustainable%20Coastal%20Livelihoods%20-%20IUCN%20(Oct%202004).pdf

IUCN—The World Conservation Union (2000). *The IUCN Policy Statement on Sustainable Use of Wild Living Resources.* http://www.iucn.org/themes/ssc/susg/docs/policy_en.pdf

IUCN—The World Conservation Union (2003). *Shark Finning.* An IUCN information paper. http://www.flmnh.ufl.edu/fish/organizations/ssg/iucnsharkfinningfinal.pdf.

IUCN—The World Conservation Union. (2006). *2006 IUCN Red List of Threatened Species.* http://www.iucnredlist.org/

Jachmann, H., and Billiouw, M. (1997). Elephant poaching and law enforcement in the central Luangwa Valley, Zambia. *Journal of Applied Ecology,* 34, 233–244.

Jennings, S., Reynolds, J.D., and Mills, S.C. (1998). Life history correlates of responses to fisheries exploitation. *Proceedings of the Royal Society of London Series B-Biological Sciences,* 265, 333–339.

Jepson, P. (2002). The need for a better understanding of context when applying CITES regulations: The case of an Indonesian parrot—Tanimbar Corella. In Oldfield, S., ed. *The Trade in Wildlife: Regulation for Conservation*, pp. 153–162. Earthscan: London.

Jett, D.A., and Nichols, J.D. (1987). A field comparison of nested grid and trapping web density estimators. *Journal of Mammalogy*, 68, 888–892.

Job Monkey (2005). *Alaska Halibut Jobs*. http://www.jobmonkey.com/alaska/html/halibut.html

Jolly, G.M. (1965). Explicit estimates from capture-recapture data with both death and immigration—Stochastic model. *Biometrika*, 52, 225–247.

Jones, J.P.G., Andriahajaina, F.B., Hockley, N.J., Balmford, A.P., and Ravoahangimalala, O.R. (2005). A multidisciplinary approach to assessing the sustainability of freshwater crayfish harvesting in Madagascar. *Conservation Biology*, 19, 1863–1871.

Jones, J.P.G., and Coulson, T. (2006). Population regulation and demography in a harvested freshwater crayfish from Madagascar. *Oikos*, 112, 602–611.

Jonzen, N., Ripa, J., and Lundberg, P. (2002). A theory of stochastic harvesting in stochastic environments. *American Naturalist*, 159, 427–437.

Joseph, L.N., Field, S.A., Wilcox, C., and Possingham, H.P. (2006). Presence-absence versus abundance data for monitoring threatened species. *Conservation Biology*, 20, 1679–1687.

Kalpers, J., Williamson, E.A., Robbins, M.M., McNeilage, A., Nzamurambaho, A., Lola, N., et al. (2003). Gorillas in the crossfire: Population dynamics of the Virunga mountain gorillas over the past three decades. *Oryx*, 37, 326–337.

Karanth, K.U., and Nichols, J.D. (1998). Estimation of tiger densities in India using photographic captures and recaptures. *Ecology*, 79, 2852–2862.

Katsukawa, T., Matsuda, H., and Matsumiya, Y. (2002). Population reproductive potential: Evaluation of long-term stock productivity. *Fisheries Sciences*, 68, 1104–1110.

Katzner, T.E., Bragin, E.A., and Milner-Gulland, E.J. (2006). Modelling populations of long-lived birds of prey for conservation: A study of Imperial eagles in Kazakhstan. *Biological Conservation*, 132, 322–335.

Katzner, T.E., Milner-Gulland, E.J., and Bragin, E.A. (2007). Using modelling to assess and improve monitoring: Are we collecting the right data? *Conservation Biology*, 21, 241–252.

Keating, K.A., and Cherry, S. (2004). Use and interpretation of logistic regression in habitat selection studies. *Journal of Wildlife Management*, 68, 774–789.

Kell, L.T., Pilling, G.M., Kirkwood, G.P., Pastoors, M., Mesnil, B., Korsbrekke, K., et al. (2005). An evaluation of the implicit management procedure used for some ICES roundfish stocks. *ICES Journal of Marine Science*, 62, 750–759.

Kellert, S.R., Mehta, J.N., and Ebbin, S.A., (2000). Community natural resource management: Promise, rhetoric, and reality. *Society & Natural Resources*, 13, 705–715.

Kerr, S., Newell, R.G., and Sanchirico, J.N. (2003). *Evaluating the New Zealand Individual Transferable Quota Market for Fisheries Management*. Motu Working Paper 2003-02. http://www.motu.org.nz/motu_wp_2003_02.htm

Kirkwood, G.P., and Smith, A.D.M. (1996). Assessing the precautionary nature of fisheries management strategies. In *FAO Fisheries Technical Paper*, No. 350, Part 2. Rome, FAO. http://www.fao.org/DOCREP/003/W1238E/W1238E07.htm

Kiss, A. (2004). Is community-based ecotourism a good use of biodiversity conservation funds? *Trends in Ecology and Evolution*, 19, 232–237.

Kremer, J.N. (1983). Ecological implications of parameter uncertainty in stochastic simulations. *Ecological Modelling*, 18, 187–207.

Laird, S.A. (1999). Trees, forests and sacred groves. *The Overstory*, 93. http://www.agroforestry.net/overstory/overstory93.html

Lancia, R.A., Bishir, J., Conner, M.C., and Rosenberry, S. (1996). Use of catch-effort to estimate population size. *Wildlife Society Bulletin*, 24, 731–737.

Latacz-Lohmann, U., and Schilizzi, S. (2005). *Auctions for Conservation Contracts: A Review of the Theoretical and Empirical Literature*. Report to the Scottish Executive Environment and Rural Affairs Department (Project No: UKL/001/05). ISBN-0-7559-1327-2 (Web only publication) http://www.scotland.gov.uk/Publications/2006/02/21152441/0

Laurance, W.F., Albernaz, A.K.M., Fearnside, P.M., Vasconcelos, H.L., and Ferreira, L.V. (2004). Deforestation in Amazonia. *Science*, 304, 1109–1111.

Leader-Williams, N., and Milner-Gulland, E.J. (1993). Policies for the enforcement of wildlife laws: The balance between detection and penalties in Luangwa Valley, Zambia. *Conservation Biology*, 7, 611–617.

Lebreton, J.D., Hines, J.E., Pradel, R., Nichols, J.D., and Spendelow, J.A. (2003). Estimation by capture-recapture of recruitment and dispersal over several sites. *Oikos*, 101, 253–264.

Lee, K.N. (1999). Appraising adaptive management. *Conservation Ecology*, 3(2), 3. http://www.consecol.org/vol3/iss2/art3/

Lee, R.J., Gorog, A.J., Dwiyahreni, A., Siwu, S., Riley, J., Alexander, H., *et al.* (2005). Wildlife trade and implications for law enforcement in Indonesia: A case study from North Sulawesi. *Biological Conservation*, 123, 477–488.

Lewison, R.L., Crowder, L.B., Read, A.J., and Freeman, S.A. (2004). Understanding impacts of fisheries bycatch on marine megafauna. *Trends in Ecology & Evolution*, 19, 598–604.

Ling, S., and Milner-Gulland, E.J. (in press). When does spatial structure matter in models of wildlife harvesting? *Journal of Applied Ecology*.

Lucas, H., and Cornwall, A. (2003). *Researching Social Policy*. Institute for Development Studies Working Paper 185. http://www.ids.ac.uk/ids/bookshop/wp/wp185.pdf

Lukacs, P.M., Anderson, D.R., and Burnham, K.P. (2005). Evaluation of trapping-web designs. *Wildlife Research*, 32, 103–110.

MacKenzie, D.I., Nichols, J.D., Lachman, G.B., Droege, S., Royle, J.A., and Langtimm, C.A. (2002). Estimating site occupancy rates when detection probabilities are less than one. *Ecology*, 83, 2248–2255.

MacKenzie, D.I., Nichols, J.D., Pollock, K.H., Royle, J.A., Bailey, L.L., and Hines, J.E. (2005). *Occupancy Estimation and Modeling: Inferring Patterns and Dynamics of Species Occurence*. Academic Press: Burlington, MA.

MacKenzie, D.I., and Royle, J.A. (2005). Designing occupancy studies: General advice and allocating survey effort. *Journal of Applied Ecology*, 42, 1005–1114.

Maclulich, D.A. (1937). *Fluctuations in Numbers of the Varying Hare (Lepus americanus)*. University of Toronto Studies, Biological Series, No. 43, pp. 1–136.

Manly, B.F.J. (1997). *Randomization, Bootstrap and Monte Carlo Methods in Biology*, Second Edition. CRC Press: Boca Raton, FL.

Marcot, B.G., Holthausen, R.S., Raphael, M.G., Rowland, M., and Wisdom, M. (2001). Using Bayesian belief networks to evaluate fish and wildlife population viability under land management alternatives from an environmental impact statement. *Forest Ecology and Management*, 153(1–3), 29–42. http://www.plexusowls.com/PDFs/marcot_bbns_2001.pdf

Margoluis, R., and Salafsky, N. (1998). *Measures of Success: Designing, Managing, and Monitoring Conservation and Development Projects*. Island Press: Washington, DC.

Martin, E.B., and Martin, C. (1982). *Run Rhino Run*. Chatto & Windus: London.

Martin, K. (1991). Experimental evaluation of age, body size, and experience in determining territory ownership in willow ptarmigan. *Canadian Journal of Zoology-Revue Canadienne De Zoologie*, 69, 1834–1841.

Martin, K.H., Stehn, R.A., and Richmond, M.E. (1976). Reliability of placental scar counts in the prairie vole. *Journal of Wildlife Management*, 40, 264–271.

Maxwell, S., and Frankenberger, T.R., eds. (1992). *Household Food Security: Concepts, Indicators, Measurements: A Technical Review*. Rome, Italy: UNICEF/IFAD.

Mbora, D.N.M., and Meikle, D.B. (2004). Forest fragmentation and the distribution, abundance and conservation of the Tana River red colobus (*Procolobus rufomitratus*). *Biological Conservation*, 118, 67–77.

McAllister, M.K., and Kirkwood, G.P. (1998). Bayesian stock assessment: A review and example application using the logistic model. *ICES Journal of Marine Science*, 55, 1031–1060.

McAllister, M.K., Starr, P.J., Restrepo, V.R., and Kirkwood, G.P. (1999). Formulating quantitative methods to evaluate fishery-management systems: What fishery processes should be modelled and what trade-offs should be made? *ICES Journal of Marine Science*, 56, 900–916.

McCarthy, M.A., Andelman, S.J., and Possingham, H.P. (2003). Reliability of relative predictions in population viability analysis. *Conservation Biology*, 17, 982–989.

McCarthy, M.A., and Parris, K.M. (2004). Clarifying the effect of toe clipping on frogs with Bayesian statistics. *Journal of Applied Ecology*, 41, 780–786.

McClanahan, T.R., Maina, J., and Davies, J. (2005). Perceptions of resource users and managers towards fisheries management options in Kenyan coral reefs. *Fisheries Management and Ecology*, 12, 105–112.

McClanahan, T.R., and Mangi, S.C. (2004). Gear-based management of a tropical artisanal fishery based on species selectivity and capture size. *Fisheries Management and Ecology*, 11, 51–60.

Mehta, L., Leach, M., Newell, P., Scoones, I., Sivaramakrishnan, K., and Way, S-A. (1999). *Exploring Understandings of Institutions and Uncertainty: New Directions in Natural Resource Management.* Institute for Development Studies Discussion Paper 372: Brighton, UK. http://www.ids.ac.uk/ids/bookshop/dp/dp372.pdf

Menu, S., Gauthier, G., and Reed, A. (2002). Changes in survival rates and population dynamics of greater snow geese over a 30-year period: Implications for hunting regulations. *Journal of Applied Ecology*, 39, 91–102.

Mesterton-Gibbons, M., and Milner-Gulland, E.J. (1998). On the strategic stability of monitoring: Implications for cooperative wildlife management programmes in Africa. *Proceedings of the Royal Society, B*, 265, 1237–1244.

Mills, L.S., Doak, D.F., and Wisdom, M.J. (1999). Reliability of conservation actions based on elasticity analysis of matrix models. *Conservation Biology*, 13, 815–829.

Milner, J.M., Nilsen, E.B., and Andreassen, H.P. (2007). Demographic side effects of selective hunting in ungulates and carnivores. *Conservation Biology*, 21, 36–47.

Milner-Gulland, E.J. (1993). An econometric analysis of consumer demand for ivory and rhino horn. *Environmental and Resource Economics*, 3, 73–95.

Milner-Gulland, E.J. (1994). A population model for the management of the saiga antelope. *Journal of Applied Ecology*, 31, 25–39.

Milner-Gulland, E.J. (2001). Assessing sustainability of hunting: Insights from bioeconomic modeling. In Bakarr, M.I., da Fonseca, G.A.B., Mittermeier, R., Rylands, A.B., and Painemilla, K.W., eds. *Hunting and Bushmeat Utilization in the African Rainforest*, Volume 2, pp. 113–151. Centre for Applied Biodiversity Science, Conservation International: Washington, DC. www.iccs.org.uk

Milner-Gulland, E.J., and Akcakaya, H.R. (2001). Sustainability indices for exploited populations. *Trends in Ecology & Evolution*, 16, 686–692.

Milner-Gulland, E.J., Bennett, E.L., and the SCB 2002 Annual Conference Wild Meat Group (2003a). Wild meat—The bigger picture. *Trends in Ecology and Evolution*, 18, 351–357.

Milner-Gulland, E.J., Bukreeva, O.M., Coulson, T.N., Lushchekina, A.A., Kholodova, M.V., Bekenov, A.B., *et al.* (2003b). Reproductive collapse in saiga antelope harems. *Nature*, 422, 135.

Milner-Gulland, E.J., and Clayton, L.M. (2002). The trade in wild pigs in North Sulawesi, Indonesia. *Ecological Economics*, 42, 165–183.

Milner-Gulland, E.J., Coulson, T.N., and Clutton-Brock, T.H. (2004). Sex differences and data quality as determinants of income from hunting red deer. *Wildlife Biology*, 10, 187–201.

Milner-Gulland, E.J., and Leader-Williams, N. (1992). A model of incentives for the illegal exploitation of black rhinos and elephants: Poaching pays in Luangwa Valley, Zambia. *Journal of Applied Ecology*, 29, 388–401.

Milner-Gulland, E.J., and Mace, R. (1998). *Conservation of Biological Resources*. Blackwell Science: Oxford.

Moller, H., Berkes, F., Lyver, P.O., and Kislalioglu, M. (2004). Combining science and traditional ecological knowledge: Monitoring populations for co-management. *Ecology and Society*, 9(3), 2. http://www.ecologyandsociety.org/vol9/iss3/art2/

Molloy, P.P., Goodwin, N.B., Cote, I.M., Gage, M.J.G., and Reynolds, J.D. (2007). Predicting the effects of exploitation on male-first sex-changing fish. *Animal Conservation*, 10, 30–38.

Morris, W.F., Tuljapurkar, S., Haridas, C.V., Menges, E.S., Horvitz, C.C., and Pfister, C.A. (2006). Sensitivity of the population growth rate to demographic variability within and between phases of the disturbance cycle. *Ecology Letters*, 9, 1331–1341.

Moustakas, A., Silvert, W., and Dimitromanolakis, A. (2006). A spatially explicit learning model of migratory fish and fishers for evaluating closed areas. *Ecological Modelling*, 192, 245–258.

Muchaal, P.K., and Ngandjui, G. (1999). Impact of village hunting on wildlife populations in the western Dja reserve, Cameroon. *Conservation Biology*, 13, 385–396.

Newing, H.S. (1994). Behavioural Ecology of Duikers (Cephalophus spp.) in Forest and Secondary Growth, Taï, Côte d'Ivoire. PhD thesis, University of Stirling.

Nichols, J.D., Sauer, J.R., Pollock, K.H., and Hestbeck, J.B. (1992). Estimating transition probabilities for stage-based population projection matrices using capture-recapture data. *Ecology*, 73, 306–312.

NMFS (1994). *Annual Report to Congress Regarding Administration of the Marine Mammal Protection Act*. Office of Protected Resources, National Marine Fisheries Service, Silver Springs, MD.

Norton-Griffiths, M., and Southey, C. (1995). The opportunity costs of biodiversity conservation in Kenya. *Ecological Economics*, 12, 125–139.

Novaro, A.J., Funes, M.C., and Walker, R.S. (2005). An empirical test of source-sink dynamics induced by hunting. *Journal of Applied Ecology*, 42, 910–920.

Nybakk, K., Kjelvik, A., Kvam, T., Overskaug, K., and Sunde, P. (2002). Mortality of semidomestic reindeer Rangifer tarandus in central Norway. *Wildlife Biology*, 8, 63–68.

O'Brien, S., Emahalala, E.R., Beard, V., Rakotondrainy, R.M., Reid, R.M., Raharisoa, V., *et al.* (2003). Decline of the Madagascar radiated tortoise *Geochelone radiata* due to overexploitation. *Oryx*, 37, 338–343.

Oates, J.F. (1995). The dangers of conservation by rural development: A case-study from the forests of Nigeria. *Oryx*, 29, 115–122.

Oates, J.F. (1999). *Myth and Reality in the Rain Forest: How Conservation Strategies are Failing in West Africa*. California University Press: Berkeley, CA.

O'Brien, S., Robert, B., and Tiandry, H. (2005). Consequences of violating the recapture duration assumption of mark-recapture models: A test using simulated and empirical data from an endangered tortoise population. *Journal of Applied Ecology*, 42, 1096–1104.

Olsen, K.B., Ekwoge, H., Ongie, R.M., Acworth, J., O'Kah, E.M., and Tako, C. (2001). *A Community Wildlife Management Model from Mount Cameroon*. Rural Development Forestry Network, Network paper 25e. Overseas Development Institute, London. http://www.odifpeg.org.uk/publications/rdfn/25/rdfn-25e-ii.pdf

Ostrom, E. (1990). *Governing the Commons: The Evolution of Institutions for Collective Action*. Cambridge University Press: Cambridge.

Ostrom, E., Burger, J., Field, C.B., Norgaard, R., and Policansky, D. (1999). Revisiting the commons: Local lessons, global challenges. *Science*, 284, 278–282.

Otis, D.L. (1980). An extension of the change-in-ratio method. *Biometrics*, 36, 141–147.

Owen-Smith, N. (2006). Demographic determination of the shape of density dependence for three African ungulate populations. *Ecological Monographs*, 76, 93–109.

Pandolfi, J.M., Bradbury, R.H., Sala, E., Hughes, T.P., Bjorndal, K.A., Cooke, R.G., *et al.* (2001). Global trajectories of the long-term decline of coral reef ecosystems. *Science*, 301, 955–958.

Parma, A.M., and the NCEAS Working Group on Population Management (1998). What can adaptive management do for our fish, forests, food and biodiversity? *Integrative Biology, Issues, News, and Reviews*, 1, 16–26.

Patton, M.Q. (1990). *Qualitative Evaluation and Research Methods*, Second Edition. Sage Publications: Newbury Park, CA.

Paulik, G.J., and Robson, D.S. (1969). Statistical calculations for change in ratio estimators of population parameters. *Journal of Wildlife Management*, 33, 1–27.

Pauly, D., Christensen, V., Dalsgaard, J., Froese, R., and Torres, F. (1998). Fishing don marine food webs. *Science*, 279, 860–863.

Pauly, D., and Watson, R. (2005). Background and interpretation of the 'Marine Trophic Index' as a measure of biodiversity. *Philosophical Transactions of the Royal Society of London Series B*, 360, 415–423.

Pelletier, D., and Mahevas, S. (2005). Spatially explicit fisheries simulation models for policy evaluation. *Fish and Fisheries*, 6, 307–349. http://www.ifremer.fr/isis-fish/publicationsen.php#1

Pelletier, F., Clutton-Brock, T., Pemberton, J., Tuljapurkar, S., and Coulson, T. (2007). The evolutionary demography of ecological change: Linking trait variation and population growth. *Science*, 315, 1571–1574.

Peres, C.A., and Nascimento, H.S. (2006). Impact of game hunting by the Kayapo of southeastern Amazonia: Implications for wildlife conservation in tropical forest indigenous reserves. *Biodiversity and Conservation*, 15, 2627–2653.

Peterman, R.M. (2004). Possible solutions to some challenges facing fisheries scientists and managers. *ICES Journal of Marine Science*, 61, 1331–1343.

Peters, R.H. (1983). *The Ecological Implications of Body Size*. Cambridge University Press: Cambridge.

Petit, E., and Valiere, N. (2006). Estimating population size with noninvasive capture-mark-recapture data. *Conservation Biology*, 20, 1062–1073.

Piron, L-H., and Evans, A. (2004). *Politics and the PRSP Approach: Synthesis Paper*. Overseas Development Institute Working Paper 237. http://www.odi.org.uk/prspsynthesis/index.html

Platten, J.R., Tibbetts, I.R., and Sheaves, M.J. (2002). The effect of increased line-fishing mortality on the sex ratio and the age of sex reversal of the Venus tusk fish. *Journal of Fish Biology*, 60, 301–318.

Plumptre, A.J. (2000). Monitoring mammal populations with line transect techniques in African forests. *Journal of Applied Ecology*, 37, 356–368.

Plumptre, A.J., and Harris, S. (1995). Estimating the biomass of large mammalian herbivores in a tropical montane forest—A method of fecal counting that avoids assuming a steady-state system. *Journal of Applied Ecology*, 32, 111–120.

Pollock, K.H. (1982). A capture-recapture design robust to unequal probability of capture. *Journal of Wildlife Management*, 51, 502–510.

Pollock, K.H., Conroy, M.J., and Hearn, W.S. (1995). Separation of hunting and natural mortality using ring-return models: An overview. *Journal of Applied Statistics*, 22, 557–566.

Pollock, K.H., Nichols, J.D., Brownie, C., and Hines, J.E. (1990). Statistical inference for capture-recapture experiments. *Wildlife Monographs*, 107, 1–97.

Pomeroy, R.S., and Rivera-Guieb, R. (2006). *Fishery Co-Management: A Practical Handbook.* International Development Research Centre and CABI Publishing. http://www.idrc.ca/openebooks/184-1/

Prescott-Allen, R., and Prescott-Allen, C. (1996). *Assessing the Sustainability of Uses of Wild Species.* IUCN SSC Occasional Paper 12. World Conservation Union, Gland, Switzerland.

Primack, R. (2006). *Essentials of Conservation Biology*, Fourth Edition. Sinauer.

Prince, P.A., Weimerskirch, H., Huin, N., and Rodwell, S. (1997). Molt, maturation of plumage and ageing in the Wandering Albatross. *Condor*, 99, 58–72.

Punt, A.E., and Smith, A.D.M. (2001). The gospel of maximum sustainable yield in fisheries management: Birth, crucifixion and reincarnation. In Reynolds, J.D., Mace, G.M., Redford, K.H., and Robinson, J.G., eds. *Conservation of Exploited Species*, pp. 41–66. Cambridge University Press: Cambridge.

Redford, K. (1990). The ecologically noble savage. *Orion*, 9, 24–29.

Rifflart, R., Marchand, F., Rivot, E., and Bagliniere, J.L. (2006). Scale reading validation for estimating age from tagged fish recapture in a brown trout (*Salmo trutta*) population. *Fisheries Research*, 78, 380–384.

Rivalan, P., Godfrey, M.H., Prevot-Julliard, A.C., and Girondot, M. (2005). Maximum likelihood estimates of tag loss in leatherback sea turtles. *Journal of Wildlife Management*, 69, 540–548.

Roberts, C.M. (1997). Ecological advice for the global fisheries crisis. *Trends in Ecology and Evolution*, 12, 35–38.

Robinson, J.G., and Bodmer, R.E. (1999). Towards wildlife management in tropical forests. *Journal of Wildlife Management*, 63, 1–13.

Robinson, J.G., and Redford, K.H. (1991). Sustainable harvest of neo-tropical mammals. In Robinson, J.G., and Redford, K.H., eds. *Neo-Tropical Wildlife Use and Conservation*, pp. 415–429. Chicago University Press: Chicago, IL.

Roldan, A.I., and Simonetti, J.A. (2001). Plant-mammal interactions in tropical Bolivian forests with different hunting pressures. *Conservation Biology*, 15, 617–623.

Romero, C., and Andrade, G.I. (2004). ICO approaches to tropical forest conservation—Response. *Conservation Biology*, 18, 1454–1455.

Roughgarden, J., and Smith, F. (1996). Why fisheries collapse and what to do about it. *Proceedings of the National Academy of Sciences*, 93, 5078–5083.

Rowcliffe, J.M., Cowlishaw, G., and Long, J. (2003). A model of human hunting impacts in multi-prey communities. *Journal of Applied Ecology*, 40, 872–889.

Rowcliffe, J.M., de Merode, E., and Cowlishaw, G. (2004). Do wildlife laws work? Species protection and the application of a prey choice model to poaching decisions. *Proceedings of the Royal Society of London Series B-Biological Sciences*, 271, 2631–2636.

Royle, J.A. (2004). N-Mixture models for estimating population size from spatially replicated counts. *Biometrics*, 60, 108–115.

Royle, J.A., and Nichols, J.D. (2003). Estimating abundance from repeated presence-absence data or point counts. *Ecology*, 84, 777–790.

Royle, J.A., Nichols, J.D., and Kéry, M. (2005). Modelling occurrence and abundance of species when detection is imperfect. *Oikos*, 110, 353–359.

Rudnick, J.A., Katzner, T.E., Bragin, E.A., Rhodes, O.E., and Dewoody, J.A. (2005). Using naturally shed feathers for individual identification, genetic parentage analyses, and population monitoring in an endangered Eastern imperial eagle (*Aquila heliaca*) population from Kazakhstan. *Molecular Ecology*, 14, 2959–2967.

Ruitenbeek, H.J. (2001). The great Canadian fishery collapse: Some policy lessons. *Ecological Economics*, 19, 103–106.

Rupp, S.P., Ballard, W.B., and Wallace, M.C. (2000). A nationwide evaluation of deer hunter harvest survey techniques. *Wildlife Society Bulletin*, 28, 570–578.

Saenz-Arroyo, A., Roberts, C.M., Torre, J., Carino-Olvera, M., and Enriquez-Andrade, R.R. (2005). Rapidly shifting environmental baselines among fishers of the Gulf of California. *Proceedings of the Royal Society B*, 272, 1957–1962.

Salafsky, N., Cauley, H., Balachander, G., Cordes, B., Parks, J., Margoluis, C., *et al.* (2001). A systematic test of an enterprise strategy for community-based biodiversity conservation. *Conservation Biology*, 15, 1585–1595.

Salafsky, N., Dugelby, B.L., Terborgh, J.W. (1993). Can extractive reserves save the rain forest? An ecological & socioeconomic comparison of nontimber forest product extraction systems in Peten Guatemala, and West Kalimantan, Indonesia. *Conservation Biology*, 7, 39–52.

Salafsky, N., and Margoluis, R. (1999). Threat reduction assessment: A practical and cost-effective approach to evaluating conservation and development projects. *Conservation Biology*, 13, 830–841.

Sale, P.F., Cowen, R.K., Danilowicz, B.S., Jones, G.P., Kritzer, J.P., Lindeman, K.C., *et al.* (2005). Critical science gaps impede use of no-take fishery reserves. *Trends in Ecology & Evolution*, 20, 74–80.

Sample, V.A., and Sedjo, R.A. (1996). Sustainability in forest management: An evolving concept. *International Advances in Economic Research*, 2, 165–173.

Sanchirico, J.N., and Wilen, J.E. (1999). Bioeconomics of spatial exploitation in a patchy environment. *Journal of Environmental Economics and Management*, 37, 129–150.

Sartorius, R. (1996). The third generation logical framework approach: Dynamic management for agricultural research projects. *The Journal of Agricultural Education and Extension*, 2, 49–62. http://library.wur.nl/ejae/v2n4-6.html

SCOS (2006). *Scientific Advice on Matters Relating to the Management of Seal Populations: 2006*. NERC Special Committee on Seals. http://smub.st-and.ac.uk/CurrentResearch.htm/SCOS2006/SCOS%202006%20collated%20document%20FINAL.pdf.

Seber, G.A.F. (1965). A note on the multiple-recapture census. *Biometrika*, 52, 249–259.

Seber, G.A.F. (1982). *The Estimation of Animal Abundance and Related Parameters*. Macmillan: New York, NY.

Seber, G.A.F. (1986). A review of estimating animal abundance. *Biometrics*, 42, 267–292.

Seber, G.A.F. (1992). A review of estimating animal abundance, 2. *International Statistical Review*, 60, 129–166.

Seidel, J.V. (1998). *Qualitative Data Analysis*. Appendix E of The Ethnograph Manual, version 5. Qualis Research. www.qualisresearch.com.

Sekhran, N. (1996). *Pursuing the 'D' in Integrated Conservation and Development Projects (ICADPs): Issues and Challenges for Papua New Guinea*. Rural Development Forestry Network, Network Paper 19b. http://www.odifpeg.org.uk/publications/rdfn/19/b.html

Selous, F.C. (1908). *African Nature Notes and Reminiscences*. Macmillan: London.

Sen, A. (1981). *Poverty and Famines. An Essay on Entitlement and Deprivation*. Clarendon Press: Oxford.

Shea, K., Possingham, H.P., Murdoch, W.W., and Roush, R. (2002). Active adaptive management in insect pest and weed control: Intervention with a plan for learning. *Ecological Applications*, 12, 927–936. http://eprint.uq.edu.au/archive/00003499/

Shea, K., Wolf, N., and Mangel, M. (2006). Influence of density dependence on the detection of trends in unobserved life-history stages. *Journal of Zoology*, 269, 442–450.

Sibly, R.M., Barker, D., Denham, M.C., Hone, J., and Pagel, M. (2005). On the regulation of mammals, birds, fish, and insects. *Science*, 309, 607–610.

Sievanen, L., Crawford, B., Pollnac, R., and Lowe, C. (2005). Weeding through assumptions of livelihood approaches in ICM: Seaweed farming in the Philippines and Indonesia. *Ocean and Coastal Management*, 48, 297–313.

Silva-Matos, D.M., Freckleton, R.P., and Watkinson, A.R. (1999). The role of density dependence in the population dynamics of a tropical palm. *Ecology*, 80, 2635–2650.

Sinclair, A.R.E., Fryxell, J., and Caughley, G. (2005). *Wildlife Ecology, Conservation and Management*. Blackwell Science: Oxford.

Sinha, A., and Bawa, K.S. (2002). Harvesting techniques, hemiparasites and fruit production in two non-timber forest tree species in south India. *Forest Ecology and Management*, 168, 289–300.

Sirén, A.H., Cardenas, J.C., and Machoa, J.D. (2006). The relation between income and hunting in tropical forests: An economic experiment in the field. *Ecology and Society*, 11(1), 44. http://www.ecologyandsociety.org/vol11/iss1/art44/

Skalski, J.R., Millspaugh, J.J., and Spencer, R.D. (2005a). Population estimation and biases in paintball mark-resight surveys of elk. *Journal of Wildlife Management*, 69, 1043–1052.

Skalski, J.R., Ryding, K., and Millspaugh, J.J. (2005b). *Wildlife Demography: Analysis of Sex, Age and Count Data*. Elsevier/Academic Press: Amsterdam.

Skonhoft, A., Yoccoz, N.G., Stenseth, N.C., Gaillard, J.M., and Loison, A. (2002). Management of chamois (*Rupicapra rupicapra*) moving between a protected core area and a hunting area. *Ecological Applications*, 12, 1199–1211.

Slade, N.A., Gomulkiewicz, R., and Alexander, H.M. (1998). Alternatives to Robinson and Redford's method of assessing overharvest from incomplete demographic data. *Conservation Biology*, 12, 148–155.

Smith, M.D., and Wilen, J.E. (2003). Economic impacts of marine reserves: The importance of spatial behaviour. *Journal of Environmental Economics and Management*, 46, 183–206.

Smith, P.J. (2005). *Analysis of Failure and Survival Data*. CRC Press: Boca Raton, FL.

Smith, V. (1994). Economics in the laboratory. *Journal of Economic Perspectives*, 8, 113–131. http://www.ices-gmu.org/article.php/369.html

SMRU (2004). *Sea Mammal Research Unit Scientific Report 1999–2004*. http://smub.st-and.ac.uk/pdfs/SMRU_Scientific_Report.pdf

Soehartono, T., and Newton, A.C. (2002). The gaharu trade in Indonesia: Is it sustainable? *Economic Botany*, 56, 271–284.

Solberg, E.J., Saether, B.E., Strand, O., and Loison, A. (1999). Dynamics of a harvested moose population in a variable environment. *Journal of Animal Ecology*, 68, 186–204.

Solis Rivera, V., and Edwards, S.R. (1998). Cosigüina, Nicaragua: A case study in community-based management of wildlife. In Milner-Gulland, E.J., and Mace, R., eds. *Conservation of Biological Resources*, pp. 206–224. Blackwell Science: Oxford.

Spash, C., Urama, K., Burton, R., Kenyon, W., Shannon, P., and Hill, G. (in press). Motives behind willingness to pay for improving biodiversity in a water ecosystem: Economics, ethics and social psychology. *Ecological Economics*.

Stanley, T.R., and Richard, J.D. (2005). Software review: A program for testing capture-recapture data for closure. *Wildlife Society Bulletin*, 33, 782–785.

StatSoft (n.d). *Survival/Failure Time Analysis*. http://www.statsoft.com/textbook/stsurvan.html#general

Stephens, P.A., Frey-Roos, F., Arnold, W., and Sutherland, W.J. (2002). Sustainable exploitation of social species: A test and comparison of models. *Journal of Applied Ecology*, 39, 629–642.

Stiles, D (2004). The ivory trade and elephant conservation. *Environmental Conservation*, 31, 309–321.

Strauss, A., and Corbin, J. (1998). *Basics of Qualitative Research: Techniques and Procedures for Developing Grounded Theory*, Second Edition. Sage Publications: Thousand Oaks, CA.

Sumaila, U.R., Guénette, S., Alder, J., and Chuenpagdee, R. (2000). Addressing ecosystem effects of fishing using marine protected areas. *ICES Journal of Marine Science*, 57, 752–760.

Sutherland, W.J., ed. (1996). *Ecological Census Techniques*. Cambridge University Press: Cambridge.

Tanner, C. (1999). Constraints on environmental behaviour. *Journal of Environmental Psychology*, 19, 145–157.

Thompson, S.K. (2002). *Sampling*. John Wiley and Sons: New York, NY.

Thompson, W.L., ed. (2004). *Sampling Rare or Elusive Species: Concepts, Designs, and Techniques for Estimating Population Parameters*. Island Press: Washington, DC.

Thorbjarnarson, J., and Velasco, A. (1999). Economic incentives for management of Venezuelan caiman. *Conservation Biology*, 13, 397–406.

Tisen, O.B., Ahmad, S.H., Bennett, E.L., and Meredith, M.E. (1999). *Wildlife Conservation and Local Communities in Sarawak, Malaysia*. Paper presented at the second regional forum for Southeast Asia of the IUCN World Commission for Protected Areas, Pakse, Lao PDR, December 1999. http://www.mered.org.uk/mike/papers/Communities_Pakse_99.htm

Tudela, S. (2004). *Ecosystem Effects of Fishing in the Mediterranean: An Analysis of the Major Threats of Fishing Gear and Practices to Biodiversity and Marine Habitats*. Studies and Reviews. General Fisheries Commission for the Mediterranean. No. 74. Rome, FAO, 44pp. http://www.fao.org/docrep/007/y5594e/y5594e00.htm

Tyre, A.J., Possingham, H.P., and Lindenmayer, D.B. (2001). Inferring Process from Pattern: Can Territory Occupancy Provide Information about Life History Parameters? *Ecological Applications*, 11(6), 1722–1737. http://eprint.uq.edu.au/archive/00003498/

Udevitz, M.S., and Ballachey, B.E. (1998). Estimating survival rates with age-structure data. *Journal of Wildlife Management*, 62, 779–792.

Udevitz, M.S., and Pollock, K.H. (1991). Change-in-ratio estimators for populations with more than two subclasses. *Biometrics*, 47, 1531–1546.

Udevitz, M.S., and Pollock, K.H. (1995). Using effort information with change-in-ratio data for population estimation. *Biometrics*, 51, 471–481.

van der Voort, M.E., and McGraw, J.B. (2006). Effects of harvester behavior on population growth rate affects sustainability of ginseng trade. *Biological Conservation*, 130, 505–516.

van der Wal, M., and Djoh, E. (2001). *Community Hunting Zones: First Steps in the Decentralisation of Wildlife Management*. Observations from the village of Djaposten, Cameroon. Rural Development Forestry Network, Network paper 25e. Overseas Development Institute, London. http://www.odifpeg.org.uk/publications/rdfn/25/rdfn-25e-i.pdf

Vaughan, I.P., and Ormerod, S.J. (2003). Improving the quality of distribution models for conservation by addressing shortcomings in the field collection of training data. *Conservation Biology*, 17, 1601–1611.

Vaughan, I.P., and Ormerod, S.J. (2005). The continuing challenges of testing species distribution models. *Journal of Applied Ecology*, 42, 720–730.

von Bertalanffy, L. (1957). Quantitative laws in metabolism and growth. *The Quarterly Review of Biology*, 32, 217–231.

Vrije Universiteit. (2001). *External Evaluation: Mount Elgon Integrated Conservation and Development Project*. http://www.mtnforum.org/oldocs/191.doc

Wacher, T., Newby, J., Houston, W. E. S., Barmou, M., and Issa, A. (2005). *Sahelo—Saharan Interest Group Wildlife Surveys. Tin Toumma & Termit (February–March 2004)*, ZSL Conservation Report Report No. 5. Zoological Society of London, London.

Wade, P.R. (1998). Calculating limits to the allowable human-caused mortality of cetaceans and pinnipeds. *Marine Mammal Science*, 14, 1–37.

Wade, P.R. (2000). Bayesian methods in conservation biology. *Conservation Biology*, 14, 1308–1316.

Walker, B.H., and Janssen, M.A. (2002). Rangelands, pastoralists and governments: Interlinked systems of people and nature. *Philosophical Transactions of the Royal Society London: Biological Sciences*, 357, 719–725. http://www.cse.csiro.au/people/brianwalker/publications/RangeRoyal%20SocPaper.pdf

Walters, C. (1986). *Adaptive Management of Renewable Resources*. Macmillan: New York, NY.

Walters, C. (2003). Folly and fantasy in the analysis of spatial catch rate data. *Canadian Journal of Fisheries and Aquatic Sciences*, 60, 1433–1436.

Weir, J.E.S., and Corlett, R.T. (2007). How far do birds disperse seeds in the degraded tropical landscape of Hong Kong, China? *Landscape Ecology*, 22, 131–140.

Whitley, D.S. (1994). By the hunter, for the gatherer: Art, social relations and subsistence change in the Prehistoric Great Basin. *World Archaeology*, 25, 356–373.

Whitman, K., Starfield, A.M., Quadling, H.S., and Packer, C. (2004). Sustainable trophy hunting of African lions. *Nature*, 428, 175–178.

WildCRU (n.d.). *Community Conservation Education in Africa*. Wildlife Conservation Research Unit, University of Oxford, UK. http://www.wildcru.org/research/other/dramagroup.htm

Wilkie, D.S., and Godoy, R.A. (2001). Income and price elasticities of bushmeat demand in lowland Amerindian societies. *Conservation Biology*, 15, 761–769.

Wilkie, D.S., Starkey, M., Abernethy, K., Effa, E.N., Telfer, P., and Godoy, R. (2005). Role of prices and wealth in consumer demand for bushmeat in Gabon, central Africa. *Conservation Biology*, 19, 268–274.

Williams, B.K., Nichols, J.D., and Conroy, M.J. (2002). *Analysis and Management of Animal Populations*. Academic Press: San Diego, CA.

Willis, T.J. (2001). Visual census methods underestimate density and diversity of cryptic reef fishes. *Journal of Fish Biology*, 59, 1408–1411.

Willis, T.J., Millar, R.B., Babcock, R.C., and Tolimieri, N. (2003). Burdens of evidence and the benefits of marine reserves: Putting Descartes before des horse? *Environmental Conservation*, 30, 97–103.

Wilson, G.A., and Rannala, B. (2003). Bayesian inference of recent migration rates using multi-locus genotypes. *Genetics*, 163, 1177–1191.

Wilson, K.R., and Anderson, D.R. (1985a). Evaluation of a nested grid approach for estimating density. *Journal of Wildlife Management*, 49, 675–678.

Wilson, K.R., and Anderson, D.R. (1985b). Evaluation of two density estimators of small mammal population size. *Journal of Mammalogy*, 66, 13–21.

Winser, S. (2004). *Royal Geographical Society Expedition Handbook*. Profile Books: London. http://www.rgs.org/OurWork/Publications/EAC+publications/Expedition+Handbook/Expedition+Handbook.htm

Winterhalder, B., and Smith, E.A. (2000). Analyzing adaptive strategies: Human behavioral ecology at twenty-five. *Evolutionary Anthropology*, 9, 51–72.

Wisdom, M.J., Mills, L.S., and Doak, D.F. (2000). Life stage simulation analysis: Estimating vital-rate effects on population growth for conservation. *Ecology*, 81, 628–641.

Wisdom, M.J., Wales, B.C., Rowland, M.M., Raphael, M.G., Holthausen, R.S., Rich, T.D., et al. (2002). Performance of Greater Sage-Grouse models for conservation assessment in the interior Columbia Basin, USA. *Conservation Biology*, 16, 1232–1242.

Wolf, N., and Mangel, M. (2004). *Understanding the Decline of the Western Alaskan Steller Sea Lion: Assessing the Evidence Concerning Multiple Hypotheses*. Report prepared by MRAG Americas Inc., for NOAA Fisheries. http://www.soe.ucsc.edu/~msmangel/Steller%20final.pdf

World Commission on Environment and Development (1987). *Our Common Future*. Oxford Paperbacks: Oxford.

WPSI (n.d.). *Tiger Poaching and Illegal Wildlife Trade Investigations.* Wildlife Protection Society of India. http://www.wpsi-india.org/projects/poaching_investigation.php

Wright, G., and Boyd, H. (1983). Numbers, age and sex of greylag and pink-footed geese shot at Loch Leven National Nature Reserve, 1966–1981. *Wildfowl*, 34, 163–167.

Zbinden, S., and Lee, D.R. (2005). Paying for environmental services: An analysis of participation in Costa Rica's PSA program. *World Development*, 33, 255–272.

Index

Note: page numbers in *italics* refer to Figures and Tables.